DATE			

High
Technology
&
Human
Freedom

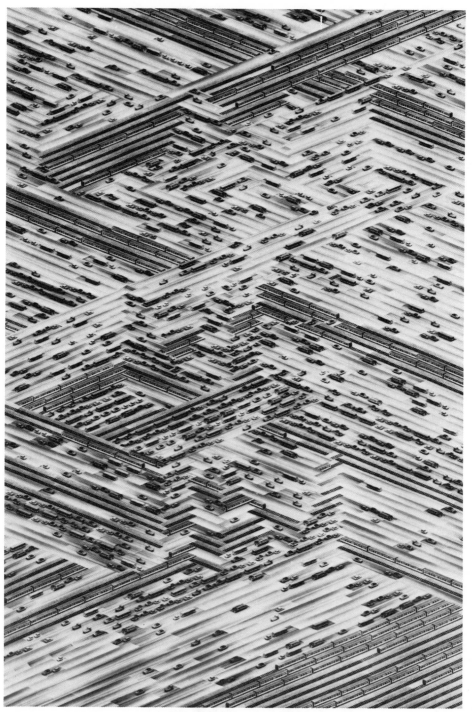

Thomas Bayrle, *Call Me Jim*

High Technology

&

Human Freedom

Edited by Lewis H. Lapham

SMITHSONIAN INSTITUTION PRESS
Washington, D.C.

List of Illustrations

© 1985 by Smithsonian Institution. All rights reserved.
Published by the Smithsonian Institution Press, Washington, D.C.

Library of Congress Cataloging-in-Publication Data

Main entry under title:

High technology and human freedom.

Papers presented at a symposium sponsored by Smithsonian Institution, Washington, D.C., Dec. 1983.
1. Technology—Philosophy—Congresses.   2. Technology—Social aspects—Congresses.   I. Lapham, Lewis H.   II. Smithsonian Institution.
T14.H46 1985   303.4'83   85-8341
ISBN 0-87474-598-5
ISBN 0-87474-599-3 (pbk.)

The paper used in this publication meets the minimum requirements of the American National Standard for Permanence of Paper for Printed Library Materials Z39.48–1984.

Book design by Christopher Jones.

Acknowledgments

The Smithsonian Institution gratefully acknowledges the cooperation and support received from the following patrons, without whose generous assistance the eighth international symposium could not have taken place:

Advanced Micro Devices, Inc.

AFL-CIO

American Bar Association Commission on Public Understanding About the Law

Benton Foundation

Centre National des Industries et des Techniques *(La Défense, France)*

Charles F. Kettering Foundation

Electronic Data Systems Corporation

Ellis L. Phillips Foundation

Exxon Education Foundation

Ford Foundation

Gannett Foundation

General Electric of Venezuela, S.A.

General Instrument Corporation

Hogan Systems, Inc.

Joseph H. Hazen Foundation, Inc.

M/A-COM Development Corporation

Nagata, Hideo and Yukiko

Nippon Electric Company, Ltd.

N.V. Philips' Gloeilampenfabrieken *(Eindhoven, The Netherlands)*

Psi Search Institute

Rosenstiel Foundation

Sage Foundation

SERI Renault Engineering *(Paris, France)*

Southern New England Telephone

Tandy Corporation/Radio Shack

United Education and Software

UVB Foundation

Winthrop Rockefeller Foundation

The symposium on which this book is based was designed and organized by the Office of Smithsonian Symposia and Seminars as part of the Institution's interdisciplinary perspective in working for the increase and diffusion of knowledge.

Wilton S. Dillon, *director*
Dorothy Richardson, *associate director*
Carla M. Borden, *associate director*
Helen Leavitt, *assistant*
Barrick W. Groom, *consultant*
Jutta Lewis and Melanie Rock, *project specialists*

Contents

Introduction

LEWIS H. LAPHAM

During the last quarter of a century George Orwell's novel *Nineteen Eighty-Four* has furnished two generations of humanists with their preferred image of a lost future. Having become synonymous with the evils of technology and the hideousness of a police state, the title stands as a metaphor that needs as little further introduction as Coca-Cola or Johnny Carson. It doesn't matter that Orwell intended the novel as a bitter satire and dark cartoon. Within a few years of its publication, in 1948, the book already had been accepted as a classic text, ·and classic texts, at least in American schools, must be treated with the deference owed to the wife of the university provost. Orwell's argument sustained the thesis of the cold war, and his bleak parable was interpreted as solemn prophecy.

As the reader will no doubt remember from English 23, or possibly Civics 104, Orwell presupposed a world divided among three totalitarian states, each of them imposing on its subject population the conditions normal to a penal colony. The food was bad; so were the conversation and the gin. Nobody was permitted to think, make love, or take notes. So powerful had these states become, so efficient and sure-handed in the techniques of oppression, that no mere individual would dare commit an act of the imagination. The reigning despots presided over bureaucracies so perfect that they could make both time and the world stand still.

Judged as prophetic description, the Orwellian portrait bears little or no resemblance to the facts. Although possibly comforting to professors of political science, the idea that the world can be governed by monolithic states has been proved laughably wrong by the events of the last twenty years. The diplomatic and military initiatives have shifted to the concentrations of power still small enough to muster a coherent expression of energy and will. Argentine generals, Central American guerrillas, African despots, Turkish or Sikh assassins, Libyan terrorists, and Israeli tank commanders—all these unlicensed individuals blithely carry on their wars and coups d'états without receiving so much as a written excuse from the faculty deans in Washington, Peking, and Moscow.

Within the past generation the Chinese have carried out at last two revolutions, and the Soviet Union, like the United States, has been hard put to preserve the semblance of empire. The Polish dock workers defied the regime, but the Politburo neglected to send troops. Within a year of his arrest Lech Walesa was giving interviews to American television.

Agents of the World Bank returning from Africa or Latin America wish they could find even a shred of evidence for Orwell's theoretical stasis. They talk instead about the chaotic entangling of the lines of international trade and credit, about the aboriginal bankrupts of the third world holding the civilized banks hostage like so many missionaries sitting in cannibal pots. In the sciences the news is the same. The correspondents on the frontiers—of physics, of biology, of astronomy, of genetics—speak of violent revolution, of fundamental changes following so fast one upon the other that nobody can conceive of a reality so small and tidy as Orwell's totalitarian housing development.

Within the United States things have gone so far beyond the bounds of despotic taste that President Reagan, quite unlike the supreme bureaucrat known as Big Brother, cannot suppress the rumors being broadcast through the White House switchboard. Prime time television glistens with the display of pornography in a variety of textures—hard, soft, wet, and suburban—and the so-called underground economy, invisible to the IRS as well as to the local police, now comprises, by conservative estimate, transactions amounting to at least $250 billion a year.

Even so, despite its failures of analysis or inaccuracies as prophecy, Orwell's *Nineteen Eighty-Four* keeps its place as the still dominant metaphor for the soul of twentieth-century government. Readers may disagree with its politics, but they cannot dismiss or ignore its poetics. Nor has anybody been able to invent so persuasive a metaphor that might cast the future in a happier and more optimistic light. Countless seers, many of them supplied with grants from the foundations or the national endowments, have attempted portraits of a postindustrial paradise in which computers take the place of kindly elves. But none of these tableaux, whether staged by Alvin Toffler, George Gilder, John Naisbitt, or the late Herman Kahn, excites the emotions so effectively as Orwell's grim anti-Utopia.

For this reason Orwell's text offered the most convenient premise for the symposium convened by the Smithsonian Institution in December 1983 in Washington, D.C., under the title "The Road after 1984: High Technology and Human Freedom." Most of the essays in the present volume were first presented as papers or speeches at that symposium. The authors all took for granted a common ground of foreboding as well as a common hope of finding a way out of the familiar maze.

During the last twenty-five years the very idea of a future has received a notoriously bad press. As preliminary to its symposium the

Smithsonian Institution commissioned Louis Harris Associates to take soundings of the public attitude about the likely effect of high technology on the structure of society. Although a majority of the respondents thought that continued scientific research probably would do "more good than harm," an impressive 69 percent of those questioned believed that Orwell's version of a totalitarian state was at least "somewhere close" to the imminent and probable facts. Within the supposedly enlightened circles of opinion, of course, it has been customary ever since the assassination of President Kennedy to speak of the future as of something dark and unclean, as if the next ten or twenty years inevitably constitute a monstrous womb certain to give birth to mutant and crawling things barely recognizable as human. The alarms have been raised, not only by the environmentalists and the proponents of disarmament, but also by the forecasters of economic catastrophe and the recruiting agents for the fundamentalist religions, in Iran, Nigeria, and Tennessee, that promise a safe return to a simple society in which the distinctions between good and evil once again appear clearly marked on the existential labels. Much of the doubt about the future reflects the suspicion that the wonders of modern technology have escaped the custody of their trainers, that maybe the latter-day sorcerers' apprentices, no matter how good their intentions or how many their Nobel Prizes, cannot put their equations back in the corner with the steam engine and the crossbow. The levels of apprehension can be inferred from the efforts of the media to domesticate the new machinery. The movies make protagonists of robots and automobiles; the magazines publish page after page of four-color advertisements in which computers display the attributes of friendly pets.

The most expensive debates in any age resolve themselves into variations on the question, Why do I have to die? As recently as the nineteenth century the question could be addressed by the literary imagination—by poets, jurists, clergymen, playwrights, and wandering moralists. The events of the twentieth century have remanded the question to the politicians and the technologists. The politicians have access to the final weapons, and the technologists will perhaps come upon the chance of immortality as well as the certainty of universal doom. If the fifteenth century saw the face of God in the Sistine ceiling, the twentieth century looks for the same consolation in a laboratory glass or on the immaculate surfaces of an ICBM.

The contributors to the present volume attempt to return the debate to the province of the humanities, to reverse the tendency expressed in Jacques Ellul's definition of modernity as "continuously improved means to carelessly defined ends." The task is one of the imagination, and maybe it is impossible to perform. Maybe it is as preposterous as Don Quixote's engagement with the windmills. Even so, if it is true that the world is something made and not something found, then the attempt must be made. In each of the following essays the prospect of fitting the new technologies to a human scale is considered. Whatever their perspective or experience, whether in law or language or history, the authors try to deal with specific instances. In so doing they know that they raise more questions than they can answer—Will the technology exert a centripetal or a centrifugal force? Within a society already seeming to dissolve into elements of special interest, must the technology be brought under the rein of the state? If every individual can set himself up as a moral entrepreneur and make a universe of a computer terminal, then who can prevent the further collapse into anarchy? Who will define the good life, and of what will it consist? Who will labor for whom, and what will be left for the schools to teach? Will it be possible to make robots into people and people into robots?

As might be expected, none of these questions admits of simple answers. Still, they must be asked, and if the essays in this book can be seen as parts of an as yet unseen whole, if they sketch the lines of connection of the newer scientific hypotheses with both their literary and their political and economic corollaries, then conceivably the road after 1984 will be a little easier to see and the metaphor of the far side of Orwell's prison a little easier to construct.

THE ROAD
AFTER 1984

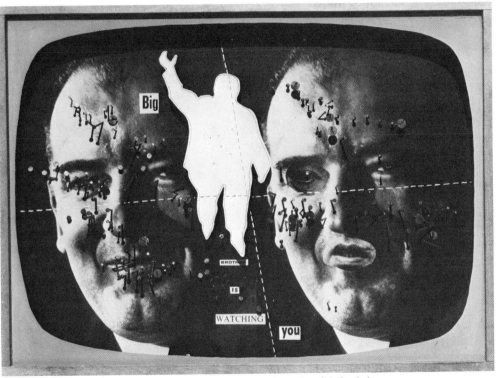

Jaap Mooy, *Big Brother Is Watching You (McCarthy)*

Thinking Machines

ROBERT WRIGHT

In July 1979, Italy's Luigi Villa, the world backgammon champion, found himself face-to-face with a robot in a $5,000 winner-take-all match at the Winter Sports Palace in Monte Carlo. The robot was linked by phone to Pittsburgh's Carnegie Mellon University, where a Digital Equipment Corporation PDP-10 computer, animated by a program called BKG 9.8, mulled things over. The human being was a 2 to 1 favorite; no machine had ever beaten a champion in a board or card game.

But BKG 9.8 beat the odds. It won four of five games and, through judicious use of the doubling die, converted that advantage into a score of 7 to 1. "Only one thing marred the scene," recalled the creator of the program, Hans Berliner, writing in *Scientific American.* "Villa, who only a day earlier had reached the summit of his backgammon career in winning the world title, was disconsolate. I told him I was sorry it had happened and that we both knew he was really the better player."

Berliner's trade is artificial intelligence, or AI. Its goal, as defined by Berliner, is to make computers do things "that if a human being were to do them, he would be considered intelligent."

Defined broadly, AI has room for two kinds of researcher. The aim of the pragmatists in the field is to replicate the results, but not necessarily the processes, of human cognition. They do not care whether their machines *think* like human beings so long as they *act* like human beings. Thus, the electronic chess boards that have brought automated defeat within reach of middle-income Americans do not win the way people win—by discerning and short-circulating the opposition's strategies, or by forging boldly ahead with a master plan of their own, or by venting their aggression move by move. These machines rely instead on superhuman feats of calculation. At each juncture they trace out thousands of possible sequences of moves and countermoves, keeping track of the pieces won and lost along the way, and then assign each possible action a number reflecting its likely long-term value. The rest even a human being could do: make the move with the highest number.

This essay originally appeared, in somewhat different form, in *The Wilson Quarterly,* Winter 1984. Copyright © 1984 by The Woodrow Wilson International Center for Scholars.

The other kind of AI researcher—the "purist"—is a programmer. Like Berliner, programmers see their mission partly as the duplication of the human thinking process. They write programs that work the way the mind works—or the way they suspect it works. To them, BKG 9.8 represents a theory of the way backgammon players think.

Whether or not programs like BKG 9.8 can be said to show intelligence, they have exhibited some facsimiles reasonable enough to impress students of human behavior. During the past decade, AI has attracted cognitive psychologists in search of a fruitful metaphor for the mind, a fresh stock of terminology, or both. They have packed psychology journals with flow charts of the human thinking process, complete with "preprocessing mechanisms," "verbal protocols," and "binary-octal translation schemes." Thus empowered, their models of the mind can "recover perceptual input" and effect "representation and retrieval of stored semantic information"—even though they sometimes labor under "incomplete feedback conditions."

As Princeton's George Miller has written, many psychologists have come to take for granted "that men and computers are merely two different species of a more abstract genus called information processing systems."

So have some journalists. The press regularly recounts the exploits of ingenious AI researchers whose progeny "think" like doctors and "understand" news articles well enough to summarize them. Alas, as computer scientists themselves concede, such accounts fall somewhere between oversimplification and distortion. *Newsweek* reporting in 1980 that computers can "draw analogies among Shakespearean plays," conjured up images of an IBM 4300 poring over *Macbeth* and then turning to a worn copy of *King Lear*. In fact, the computer in question scanned plot summaries that read more like FORTRAN than Elizabethan English: "Macbeth is a king. Macbeth marry Lady-Macbeth. Lady-Macbeth is a woman—has property greedy ambitious. Duncan is a king. Mac-duff is a noble—has property loyal angry. Weird-sisters is a hag group—has property old ugly weird—number 3."

Once the hyperbole has been stripped away, computer scientists turn out to be only human—and to consider their machines only machines. The early optimism for AI has been tempered. The difficulty of replicating even the more mundane cognitive functions has left some researchers saying what poets, mystics, and various other skeptics have said all along: the mind is not a computer—at least, not the kind of computer that has been built so far.

Artificial intelligence might well have been called cybernetics, the discipline under which scientists first tried to simulate thinking electronically. Cybernetics began during the 1940s as the study of feedback systems. Its founder, the MIT mathematician Norbert Wiener, sought to make allied antiaircraft guns self-aiming by giving them radar information about the speed and direction of the target. The parallels between this "feedback loop" and the human nervous system suggested that comparisons between mind and machine might be productive—an idea that fed on enthusiasm about newfangled "electronic computing machines." Soon cyberneticists were building networks of elaborately interconnected electronic switches, modeled after the brain's masses of neurons. But these "neural nets" were drastically oversimplified models of the mind and displayed little intelligent behavior. By the late 1960s, this line of research had reached a dead end.

The term "artificial intelligence" was coined by Stanford's John McCarthy to describe the focus of a conference held in 1956 at Dartmouth, where he then taught mathematics. In a grant proposal submitted to the Rockefeller Foundation, McCarthy, Marvin Minsky, now head of MIT's artificial intelligence laboratory, the mathematician Claude Shanon, and IBM engineer Nathaniel Rochester wrote that the meeting would deal with the "conjecture" that every feature of intelligence "can in principle be so precisely described that a machine can be made to simulate it."

In some respects, the conference supported that conjecture. Allan Newell, Herbert Simon, and J. C. Shaw of Carnegie Tech (now Carnegie-Mellon) and the Rand Corporation introduced a program called Logic Theorist, a prodigy in symbolic logic. Confronted with fifty-two of the theorems proved by Alfred North Whitehead and Bertrand Russell in *Principia Mathematica* (1925), Logic Theorist managed to prove thirty-eight of them—and one of its proofs was more "elegant"—that is, more straightforward—than the original.

This was all the more impressive because Logic Theorist did not rely on brute force. It did not simply try every possible combination of logical rules until it came across one that worked. Instead, it employed "heuristics," rules of thumb that narrow one's focus in the face of a bewildering array of options that sometimes lead nowhere; Newell, Shaw, and Simon, intent on modeling human thinking, chose to make their program fallible.

Flushed with success, Simon ambitiously staked out AI's territory. There are now, he declared, "machines that think, that learn, and that create. Moreover, their ability to do these things is going to increase rapidly until the range of problems they can handle will be coextensive with the range to which the human mind has been applied."

During the next few years, computer scientists produced one intriguing plaything after another. The Conversation Machine, built in 1959, could make passable small talk—so long as its partner communicated by typewriter keyboard and didn't stray too far from the subject of the weather. In 1961 a program written by an MIT graduate student got an A on a calculus exam. By 1962, a string quartet had performed music composed by a computer that had used rules of counterpoint formulated by the sixteenth-century Italian composer Giovanni Palestrina.

By the mid 1960s, though, the heady years were over. Impressive as the feats of AI seemed, they still paled in comparison with the daily accomplishments of the human mind. General Problem Solver, for example, a program unveiled by Newell, Simon, and Shaw amid much fanfare in 1957, proved to be capable of less than its name suggested. True, it was more of a Renaissance man than was Logic Theorist: It could handle algebra problems and logical puzzles, such as how to get three missionaries and three cannibals across a river alive using only a two-man boat. Still, these are not the kinds of skill most people associate with the word *generalist.*

The limitations of General Problem Solver suggested that intelligence cannot be boiled down to a few versatile techniques. It seemed, rather, that the human intellect depends on a large repertoire of tools, many of them useless without vast quantities of specialized knowledge. Accordingly, during the late 1960s and early 1970s many AI researchers turned their attention to knowledge engineering, the transplanting of expertise from doctors, geologists, or mechanics to "expert systems." This research would eventually produce such programs as INTERNIST-1, an aid to medical diagnosticians. In a 1983 test that involved cases drawn from the *New England Journal of Medicine,* it proved nearly as accurate as the attending physicians.

But even with the mechanization of expertise, AI still faced a formidable barrier: the common-sense problem. Computers can second-guess Bertrand Russell, play respectable chess, and diagnose soybean plant pathology with the self-assurance of a county agent, yet they have trouble comprehending "The Farmer in the Dell."

In trying to imbue computers with common sense, researchers have had to grapple with questions about logic. How large a role does it really play in human thinking? How large a role should it play in machine thinking?

Minsky believes that the average mind rarely functions with the rigor of logic. In everyday life, he says, "I suspect we use it less for solving problems than we use it for explaining the solutions to other people and—much more important—to ourselves." Computers will not truly be thinking machines, he suggests, until they can formulate vague definitions, harbor inconsistent ideas, and, on weighing evidence and finding it incomplete, jump to the nearest conclusion.

John McCarthy, Minsky's antagonist in the logic debate, believes that strictly logical computers *can* have common sense. "The only reason we have not yet succeeded in simulating every aspect of the real world is that we have been lacking a sufficiently powerful logical calculus," he declared in 1972. "I am currently working on that problem." Thirteen years later he still is, and he attributes Minsky's lack of such faith to an intellectual bias. "Minsky never liked logic," McCarthy told a reporter for *Science*.

One of Minsky's favorite illustrations of the shortcomings of logic is the "dead duck" example. Birds can fly; a duck is a bird; Joe is a duck. A computer with powers of deduction will conclude that Joe can fly. But what if Joe is a dead duck? And what about Hubert the penguin, who is also a bird but will never take wing? A child knows that neither can fly, but a computer relying on deductive logic does not.

Exceptions can be programmed into a computer, but if there are too many, it is not worth the trouble of devising the rules in the first place. The real world, Minsky argues, is riddled with exceptions, yet people cope anyway; deductive logic, therefore, must not be central to their thinking.

Researchers trying to teach machines to comprehend "natural language" (such as English) have confronted a second shortcoming of logic. Much of what human beings absorb while reading does not follow logically from what is written. A newspaper reader does not have an airtight case, for example, in concluding that an assault victim who was "treated and released" was only slightly injured. Still, such common-sense reasoning is almost always on target.

Ambiguity further complicates matters. How is a computer to know that the meanings of *flies* and *like* can change radically from one sentence (time flies like an arrow) to another (fruit flies like an apple)? Of course, context might clarify things. Is the computer at a college reunion or at an exterminator's convention?

By giving computers just such contextual information, Roger Schank, head of Yale's AI laboratory, has attacked several problems of language comprehension. Each of his scripts sets the context, providing generally safe assumptions about the way a given situation unfolds. Schank's restaurant script keeps the computer from even contemplating the possibility that *tip* refers to Gallant Prince in the seventh at Belmont and also facilitates reading between the lines; when a customer leaves a big tip, the computer is told, it probably means that he was satisfied with the service.

Whatever their value to computers, scripts have their limitations when taken as theories of human cognition. A single script contains much information, but it would take a great many of them just to get a person through the average day. Do human beings really carry around thousands of scripts and pop a new one into the mental projector every time they move from the grocery store to the sidewalk, or turn from the obituaries to the sports page—or from a story about baseball to one about basketball? Is nature, with its preference for simplicity, really likely to build brains that have to perform such a complicated juggling act? In their simplest form, theories based on scripts suggest that this is indeed so.

There are other theories of cognition that do not call for so much shuffling of information, but not all can be tested easily on conventional computers. They are more compatible with a coming generation of machines—the "massively parallel" computers, which some tout as the new wave in AI.

Take, for example, some of Minsky's theorizing. If machines are going to think like human beings, Minsky says, they must get out of the habit of defining words with mathematical precision and instead simply associate each word with a melange of other words. They must behave more like Euthyphro, the Greek prophet who could name pious and impious men but could not give Socrates a rigorous definition of piety.

"What if we built machines that weren't based on rigid definitions?" Minsky has written. "Wouldn't they just drown in paradox, equivocation, inconsistency? Relax! Most of what people 'know' already overflows with contradictions. We will survive." An "associationist" approach to defining words, he believes, will be easier with massively parallel computers.

Virtually all today's computers are based on the von Neumann architecture developed by the mathematician John von Neumann during the 1940s. A von Neumann machine is run by a central processing unit that retrieves information from the computer's memory, modifies it according to the program, and then either returns it to memory or prints it out and forgets it. Generally, such machines can do only one thing at a time.

In a machine based on parallel architecture, though, different processors work on different aspects of a problem simultaneously. Parallel computers have been around for some time, but so far none has been—well, massive. The first that will be is the Connection Machine, versions of which are being built at MIT and at Thinking Machines, Inc., in Cambridge, Massachusetts. It will have 64,000 processors, each powerful enough to be considered a capable computer in its own right. Even so, the machine will simulate only a thin slice of the mind; MIT is already planning a larger version.

In massively parallel computers, no one processor does anything very sophisticated, and none oversees the operation of the others. Intelligence is not imposed from the top down; it emerges, from the bottom up, just as collectively intelligent behavior arises in an ant colony despite its nonhierarchical structure and conspicuous lack of individual genius.

Proponents of massive parallelism view the mind as a society. Jerome Feldman of the University of Rochester writes of "winner-take-all networks," in which "coalitions" of processors continually clash. In Feldman's model, concepts are represented, not by strings of symbols, as in a von Neumann computer, but by patterns of interconnection among processors. This approach, he says, offers a way to address the issues of ambiguity and context more economically than do scripts.

Take a sentence such as "John threw a ball for charity." In the machine envisioned by Feldman, the two senses of the verb "to throw"—to hurl, and give by way of entertainment to—would live in separate processors, or nodes. Upon encountering this sentence, both nodes would seek support for their interpretations; they would try to find other words in the sentence with which they have an affinity—that is, with which they are connected.

Both would have immediate success. The *hurl* node would be wired to the node housing the corresponding sense of *ball,* a spherical object. The second sense of *throw,* to give by way of entertainment would be linked with the second sense of *ball,* a dance. (These links could have been forged in the human mind by learning and in the computer by a programmer.) Once these two links had been activated, they would try to overcome one another.

Victory would go to the majority. When each pair tried to encompass the third key member of the sentence—the swing vote—only one would succeed. The *dance* node would be connected to the charity node; charity balls are common enough to warrant that linkage. But the more conventional sense of *ball* would search in vain for an affiliation with *charity.* The host-dance coalition would now have control of the sentence and would electronically suppress any dissent.

In Feldman's model, as in models that embody scripts, knowledge of context helps. If "John threw a ball for charity" had come up in the course of a town social committee meeting, the connection already activated would have headed off any grassroots movement for a *baseball* interpretation. Thus, Feldman says, the "connectionist paradigm" offers "dynamic" scripts. They resolve ambiguity and take account of context, but they do not come in separate, bulky packages that must be constantly juggled. Instead, a script is defined by the prevailing pattern of interconnection among tiny packets of information, all of which stay put; dynamic scripts can be modified subtly or dramatically without any reshuffling of information.

Ideas bearing some resemblance to Feldman's have been around for some time. In *Psychology* (1893), William James explored the "principles of connection" in accordance with which "points" of the brain are temporarily linked by "discharges" and thoughts "appear to sprout one out of the other." Half a century later came the cyberneticists' "neural nets," designed to learn by memorizing patterns of interconnection among nodes. Because the cyberneticists' ambitions outstripped their understanding of the mind, neural nets did not live up to their billing; the von Neumann architecture was the only game in town by the 1960s, when psychologists turned for inspiration to computer science.

Almost every Psychology 101 student since then has encountered fruits of that search, in the form of textbook flow charts that trace the path of information through a mental processor and into long-term memory storage. Had massive parallelism been in vogue in the sixties, those charts might look quite different; information would be dispersed throughout a huge honeycomb, and "bits" would be processed where they reside.

And the prospect of machines behaving intelligently might not be seen as so dehumanizing. After all, no central processing bit will exert tyrannical rule over a massively parallel machine. On the contrary, the democratic behavior of the processors will be so unruly that not even the creator of a program will always be able to predict results.

Would that uncertainty reflect a certain capriciousness on the part of the machine—even, perhaps, a trace of free will? Some computer scientists appear willing to toy with that possibility. They will go so far as to call such unpredictable behavior nondeterministic.

If massive parallelism lives up to the expectations of its advocates, the question may well be asked: Were the first thirty years of AI, with their emphasis on the top-down approach to simulating intelligence, just a long detour for all the psychologists who were persuaded to climb onto the bandwagon?

Few in AI seem to think so. Whatever the value of massive parallelism as a metaphor for mind, no one contends that it can capture the entire thought process. Herbert Simon points out that regardless of how information is processed at subconscious levels, it must ultimately pass through the bottleneck of conscious attention, which is clearly a "serial," not a parallel, processor; a person can only entertain one thought at a time.

Simon does not share Minsky's and Feldman's high hopes for massive parallelism. He does agree that logic plays a limited role in human thought—he won the 1978 Nobel Prize in economics for his theory of bounded rationality, which stresses the arbitary nature of much human decisionmaking. Still, he notes, conventional computers have demonstrated their ability to simulate nonlogical processes, even if those simulations sometimes take longer than they would on parallel machines. Much enthusiasm about massive parallelism, he says, is romanticism.

There is one point, though, on which the supporters and detractors of massive parallelism agree: no matter which of AI's competing models of thought prevails, computer science will have made a lasting contribution to cognitive psychology. At the very least, computers force a theoretician to define his terms; it is difficult to turn a lot of murky thinking into a successful computer program.

This benefit was foreseen nearly four decades ago by Harvard psychologist Edwin G. Boring. He had been challenged by Norbert Wiener to describe a capacity of the brain that no machine could ever duplicate. Just contemplating that challenge, Boring found, was enlightening. It forced him to refine his ideas about the nature of intelligence.

Boring urged others to try this experiment in their heads—to pretend, in essence, that they were computer programmers trying to simulate human thought, and consider the issues that they would then confront.

In a 1946 edition of the *American Journal of Psychology,* he asked readers: "With what property must a robot be endowed by its maker in order that it may make discriminations, may react, may learn, may use symbolic processes, may have insight, may describe the nature of its own functions and processes?" Contemplating such questions, he suggested, is the way to go at the question of how the mind works. "It is a procedure that keeps us clear."

Can computers really *think?* Either the average American considers this a burning question, or scores of journalists have badly misread the average American; scarcely an article on artificial intelligence appears that doesn't begin or end on this note. But why do people find the question so fascinating? Surely not for the reason that psychologists do—not for the light the ensuing debate sheds on human cognition. No, one gathers that the layman's interest is grounded less in curiosity than in insecurity. If computers can't think, after all, how threatening can they really be? So long as we alone can distill the seven o'clock news, detect notes of sarcasm, and never forget a face, our autonomy will remain safe, removed from the reach of less subtle machines. This assumption seems to be a common source of reassurance these days.

In one sense, this assumption is obviously false. As David Burnham notes in a later chapter, it is the less arcane applications of computer technology that could breathe life into the most literally Orwellian nightmare—pervasive government surveillance. By extending the eyes, the ears, and even the arms of the law, computers of no great intellectual distinction will create the potential not only for unprecedented police efficiency, but for high-tech persecution.

In another sense, this assumption is false, but less obviously so. Even when we restrict the computers under discussion to those blessed with artificial intelligence, the question whether they can really think turns out to be only loosely related to the question whether they should worry us. A machine with no common sense and no chance of breaking into triple figures on the SAT test can nonetheless muster enough intelligence to threaten human freedom, in various senses of the term.

Expert systems, for example, are only in their infancy, and even at maturity they will lack the street smarts that a nine-year-old demonstrates in discussing an episode of *Hill Street Blues.* Nonetheless, they are fast becoming a threat to very intelligent adults. Perhaps INTERNIST 1 would not qualify to teach at Harvard Medical School, but it can probably do a better job at certain kinds of diagnosis than do some living doctors, particularly those who treat lower-income Americans. This is good news for lower-income Americans; an expert system, though expensive to build, is cheap to clone. But members of the upper middle class, notably its bright young college students, may be less enthusiastic about the mass production of expertise; notwithstanding the gift that doctors (and lawyers) have for manufacturing a demand for their services, some young would-be professionals may soon join the ranks of the technologically unemployed and gain an appreciation for a phrase that now strikes the privileged as paradoxical: freedom to work.

Perhaps that will be a good thing. Technology has traditionally heightened the political tension between blue-collar and white-collar workers. Manual laborers, the primary victims of automation, have long advocated a more nearly even distribution of the fruits of technologically driven gains in productivity, and they have not, generally speaking, got their way; rather than reduce everyone's work week by a few hours, we have let the ranks of the unemployed swell, along with the paychecks of the employed. Once doctors, tax lawyers, and engineers are able to empathize with the technicians, typists, and drafters who have met their match in machines, perhaps society will begin to see technological fallout as community property.

But even if artificial intelligence *does* shorten every work week slightly, rather than leaving most untouched while reducing some to zero, the issue of human freedom will confront us in another form: How constructively will free time be used? Computers may open new channels of creativity, but a variety of technologies, from electronic to pharmacological, will offer vicarious pleasures that lead to new avenues of self-destruction. So we may be delivered from the tyranny of forty-hour and fifty-hour work weeks only to fall under the influence of designer drugs and cheap-thrills machines.

Of course, maybe we won't. Maybe we'll spend our free time reading poetry and staging canned-goods drives for poor people. But the experience of three out of the last four decades suggests otherwise.

In the classroom, laws of economics dictate that intelligent machines will eventually become tutors, perhaps even playmates, of the young. By the time school administrators exhaust their present compulsion to invest in computers and educational software, the day will not be far off when such investments can be soundly made; the next generation of instructional technology may possess sufficient expertise, and sufficient sensitivity, not only to qualify as artificial intelligence, but to surpass teachers in educational efficiency. In fairness to teachers, the comparison should be made more precisely: A *roomful* of computers, each bringing a student along at his own pace, and perhaps by a route tailored to his cognitive idiosyncrasies, will teach a roomful of students more in an hour than a single teacher can. What is more, the roomful of computers may be the cheaper of the two options. (Not that there is much chance of ever putting an entire school on automatic pilot. The teacher-student ratio may gradually decline but is unlikely ever to reach zero.)

These educational machines will dish out congratulations for correct answers—and, it is hoped, even for unsolicited insights—and give gentle reproofs for wrongheadedness. This will naturally evoke images of pigeons in Skinner boxes and will spark debate about freedom and control. And well it should. The boundary between intellectual and moral development is never clear-cut, and there will be values, as well as ideas, embedded in the software's electrons. The question of which values, and how forcefully they are to be inculcated, should be determined consciously and, if possible, rationally.

But, it might be asked, why bother to debate such issues formally? Much of what human teachers say to their classes is value-laden, and its content is now determined almost effortlessly, by reference to local standards: Voters elect school board members who remind them of themselves; school board members hire administrators who remind them of *them*selves, and administrators hire teachers who remind them of *them*selves.

True enough. But individual, automated tutors will probably prove more enchanting, and hence more insidious, than the average teacher, who is lucky to have the divided attention of half the class at any given moment. So, if knowledge about behavioral development grows significantly—and educational software research may spur just such growth—the stage will be set for some serious modification of behavior. Schools may never become factories—they may never mold students' intellectual and moral character to exacting specifications—but we will have the option of moving public education in that direction. As usual, there will be potential for good and for ill. It might not hurt to reread *Walden II* and think about some old issues.

Damn the Absolute

Henry Steele Commager

"Damn the Absolute." —William James

It was just a quarter century ago that C. P. Snow agitated the intellectual world with his essay on the two cultures, the one scientific, the other literary. But these two cultures did not in fact require, or impose, two ways of thinking. They merely confessed different interests.

If there are indeed two cultures, the distinction is a more fundamental one: it is that between cultures which depend on the inductive habit of thinking and those which yield to the deductive habit. It is this difference which for two millenia—and overwhelmingly since the age of Francis Bacon and Isaac Newton—has distinguished almost every arena of thought and conduct. The first of these cultures takes off from a series of axiomatic assumptions that are not open to challenge; the second is grounded on such facts as can be substantiated by experiment. This difference is between those who start from absolutes in morals, philosophy, economics, politics, law, and even international relations and those who start with the findings of the laboratory or, in the absence of these, from the somewhat more dubious authority of experience.

Reliance on experiment served as the fundamental principle of that first great scientific institution, the Royal Society, whose roots were in the Oxford Experimental Club and whose charter, written by Oxford's professor of astronomy Christopher Wren, provided that it was to "support those arts and sciences which by actual experiments attempt to shape a new philosophy to perfect the old."

So, too, the Smithsonian Institution was the creature of three scientists—James Smithson, a chemist who won admission to the Royal Society at the age of twenty-two; Joseph Henry, who proposed "to found a cosmopolitan establishment to increase the sum of human knowledge, and to diffuse it to every part of the civilized world"; and John Quincy Adams, the only president to write a distinguished scientific treatise while in the White House.

Even as the Smithsonian was being founded in the closing days of 1839, Alexis de Tocqueville was penning the final pages of what proved to be the most profound book ever written on the fledgling

United States. In *Democracy in America* he concluded that "Democracy did not exalt, nor inspire, as did old world class societies. It confessed a mediocre, almost vulgar culture, one which could display neither the brilliance, the beauty, nor the grace, of the Old World. But," he added,

> We may naturally believe that it is not the singular prosperity of the few but the greater well-being of all which is most pleasing in the sight of the Creator and preserver of men. What appears to me to be man's decline is to His eye advancement; what afflicts me is acceptable to Him. A state of equality is perhaps less elevated, but is more just; and its justice constitutes its greatness and its beauty.

This passage is a striking example of what was both Tocqueville's strength and his weakness, his tendency—we might say his genius—to combine the deductive and the inductive interpretations. A large subject, this; suffice it here to remind you that Tocqueville's famous doctrine of the tyranny of the majority was not based on evidence—at least not on any evidence that he submitted—nor was his fear of the military, nor, for that matter, were his grave warnings against the baleful effects of centralization. Yet even there he was prescient beyond any other historian of his time, for he had already foreseen the potential threat of what he called the manufacturing aristocracy—what we might call the corporate economy—and already, too, the beginnings of the vulgarization of culture. On inductive grounds, his ecstatic tribute to a just society was valid enough if the greatest injustice of all, slavery, is overlooked. Tocqueville did not overlook it, but he managed to exclude it from his general observations on democracy as Jefferson and his associates managed to exclude it from the concept of equality in the Declaration of Independence.

Except for slavery, the prospects for the new republic were bright. The United States appeared to have everything going for it: limitless territory—"Land enough for our descendants to the thousandth and thousandth generation," said Jefferson; limitless resources; immunity from invasion or conquest and the ability, therefore, to forgo a large military establishment; a longer tradition of democratic self-government than any other people; and the highest standard of literacy of any people. More relevant to our concern here, the new nation was established in the age of the scientific and industrial revolutions. "What more," Jefferson asked, "is necessary to close the circle of our felicities?"

Tocqueville himself added one further consideration: Americans were a religious people (religous rather than devout); and Protestantism was, in the United States, more suited to their genius than his own Catholicism. He saw, too, that Americans had already formulated what we have come to call a civil religion, one based on experience rather than on dogma and inheritance.

Tocqueville devoted one of his chapters to an analysis of "why the Americans are more addicted to practical than theoretical science." "In a democratic country," he observed, "people are always afraid of losing their way in visionary speculations. They mistrust systems, they adhere closely to facts ... they are never inclined to rest upon any man's authority."

"They mistrust systems, they adhere to facts." Here Tocqueville was not so prescient. The South did not mistrust systems or adhere to facts. Had they done so they would not have clung so obstinately to slavery; they would not have seceded; they would not have counted on cotton being king and that the South, therefore, would emerge from a war victorious.

Nor, as it turned out, were powerful, if not dominant, forces in the North—and eventually in the nation as a whole—invariably committed to facts when they applied Darwinian biology to the economic, the social, the political world.

We come here to the great debate over the relation of man to nature, to government, and to technology; a debate that came to a dramatic climax in the year 1884—the year that saw the publication of Herbert Spencer's *The Man versus the State* and Lester Ward's *Dynamic Sociology*. It proved to be the most heated intellectual debate in our history, and inaugurated those great controversies which, from the 1880s to the 1980s, divided American statesmen, politicians, economists, social philosophers, and moralists. The division was neither abstruse nor abstract, thought it rested on abstruse and abstract considerations. It was, broadly, between those who thought government should consist of "anarchy plus the policeman," to use Spencer's phrase, and those who championed what we call the welfare state; or between those who believed that society should let nature alone work out man's destiny, according to the laissez-faire principle of natural selection, and those who insisted that all progress is a product of interference with nature and that government is the most efficient instrument of it.

This argument for laissez faire boasted ample antecedents. (Happily for the historian there are always antecedents.) There was the wicked Bernard Mandeville, in the early eighteenth century, whose *Fable of the Bees* delighted successive generations, including our own, with its profound moral principle of "private vices, public benefits." "What we call evil in this world," Mandeville wrote, "moral as well as natural, is the grand principle that makes us social creatures—the solid basis, the light and support of all trades without exception: Luxury." And he added, for the solace of our statesmen today, "though the military, too, is condemned by moralists it is necessary to society." Then there was the Reverend Thomas Malthus, who saw in unlimited population the greatest threat to the prosperity of society and who was happy to observe that "the vices of mankind are active and able ministers of depopulation; they are the precursors of the great army of destruction, and often finish the dread work themselves." There was the magisterial Edmund Burke, who could proclaim, well after Americans had won their independence and ratified their new constitution, that it was impossible for any government to survive without an established church.

Later, Herbert Spencer became the fount of this new stream of philosophical truth; he was, to use the phrase applied to Emerson, the cow from whom his followers drew their milk. It is difficult for us now, more than a century later, to understand the passionate admiration Spencer inspired and the loyalty he commanded in the United States. "His genius," wrote the historian John Fiske, "surpassed that of Aristotle as the telegraph surpassed the carrier pigeon." President Frederick Augustus Porter Barnard of Columbia University insisted, quite simply, that Spencer's was the most capacious and powerful mind of all time. It was as biologist, psychologist, and philosopher that he had his great effect. John Fiske, who considered himself a "cosmic philosopher," and William Graham Sumner were his mockingbirds. Evolution, in the Spencerian version, was, when properly understood, benign; but it had the inevitable drawback of requiring many years. Such a version of evolution was embraced with enthusiasm by men of property and of substance everywhere.

This was the crux of the matter. Social Darwinism's human progress could come only through natural causes, as did progress in the animal kingdom, and these could be counted on only if there was no interference from man. This was the concept that enabled economic and business conservatives to wrestle themselves into Tom Paine's clothes and intone his litany, "Government, like dress, is the badge of lost innocence." It was this philosophy that provided a patina of respectability to what might otherwise have been considered flirtation with anarchy.

What has heretofore been overlooked is an affinity between what we call social Darwinism and the proslavery argument—or philosophy—of the Old South. This affinity is all the more remarkable insofar as it was chiefly the South that refused to accept the religious implications of Darwinian biology. Much of this reluctance persists. What are some of the indexes of this correspondence?

First, both the proslavery argument and the precepts of social Darwinism were based on absolutes. The absolutes were, to be sure, different—scriptural in the South, scientific in the North—but the absolutes of white supremacy and of the survival of the fittest were compatible enough. Second, both appealed to science: social Darwinism to a sound and legitimate science, however misapplied, white supremacy to nonsense, or nonscience, "proving" the biological superiority of whites to blacks. Third, both sustained, whether by design or by unavoidable logic, a class system. Fourth, both rationalized the exercise of power by a ruling class, singled out by nature herself as superior—an exercise of power which served, therefore, an honorable public interest. Fifth, both were sanctioned by a higher power: slavery by God—the Bible made that clear; the primacy of an intellectual and physical elite by the inscrutable but divinely ordained operations of nature.

Consider William Graham Sumner's prescription for salvation and progress:

> We have inherited a vast number of social ills which never came from nature. They are the complicated products of all the tinkering, muddling and blundering of social doctors in the past. These producers of social quackery are now buttressed by habit, fashion, prejudice, and new quackery in political economy and social science. The greatest reform which could not be accomplished would consist in undoing their work. (1883)

What Sumner made clear was that man was the subject, not the object, of the evolutionary process. The success of that project depended in large part upon his passivity. By associating the hands-off policy with liberty, he elevated it into a moral dogma as well.

> All experience is against state regulation and in favor of liberty, and the freer the civil institutions, the more weak or mischievous state regulation is. . . . Whatever we gain will be by growth, never in the world by any reconstruction of society on the plan of some social architect. Society needs first of all to be free from meddlers, that is, to be let alone.

Turn now to Lester Ward, the most distinquished paleontologist in the country and a polymath to boot. When at the age of sixty he became a professor at Brown University, he taught only one course: "An Outline of All Knowledge." In the larger political arena he was, some claim, the leading philosopher-architect of the welfare state in the United States.

Ward taught that man was not the passive or the helpless subject of the evolutionary process but in large measure in charge of it. "Is it true," he asked, "that man shall obtain dominion of the whole world, but not of himself?" The principle of hands off, Ward argued, was both futile and incoherent. It repudiated not only all experience, but man himself. It repudiated history and, by denying any scope to the creative faculties of man, condemned the future. All civilization, Ward insisted, was the triumph of man over the blind forces of nature: "Is not civilization with all its accomplishments the result of man's not letting things alone—that is, of substituting art for nature?"

More effectively than any other, Ward confronted the great issue of positive and passive evolution versus active evolution. He had great companions. In philosophy there was William James, a man trained in medicine, and our pioneer psychologist, who insisted that the truth of an idea is not a stagnant property in it. "Truth is what happens to an idea," was James's credo, and he applied the terms *instrumentalism* and *pragmatism* to the whole vast field of social, political, and educational philosophy. In economics there was the mordant Thorstein Veblen, who, in his theory of business enterprise, stood the absolutes of industrial capitalism on their heads. There was E. A. Ross, Ward's nephew, whose *Of Sin and Society*—more timely now than in 1906, when he wrote it—claimed that the greatest sinners were not those who indulged in private vices, but those responsible for public crimes. Jane Addams and John Dewey challenged the familiar dogmas of education—dogmas as deeply rooted as those attacked elsewhere by Ward and Veblen—and turned back to Rousseau and Pestalozzi and (let us not forget) Bronson Alcott. Most impressive, and most effective, of all was the new school of social jurisprudence: Justice Holmes and Louis Brandeis, Roscoe Pound and Ernest Freund, and a large galaxy of their disciples, who did their best to vindicate not only Holmes's aphorism, "the constitution did not enact Mr. Herbert Spencer's *Social Statics*," but also his conclusion that "the life of the law has not been logic, it has been experience." All these rejected the deductive method of reasoning and endorsed the inductive, and they all relied on science and technology, not nature, to solve problems. All were relativists; all might have taken for their battle cry William James's "Damn the absolute."

From the point of view of the "natural philosopher," technology, broadly interpreted, has been largely responsible for whatever civilization we have. Great scholars such as Fernand Braudel have traced this process for us from the earliest "traces on the Rhodean shore" to modern times—man in partnership with nature, generation by generation for thousands of years, improving upon the earth and creating civilizations, both primitive and sophisticated. It was a process in which every advance, from breeding cattle or designing tools or discovering medicinal herbs or developing speech and writing, was a deliberate and often violent interference with nature. Darwinism let nature have her way. Malthusian controls—reliance on war, famine, disease, and pestilence—also let nature have her own way. But those we call Lamarckian—which Ward and his associates welcomed as the conscious inheritance of deliberately acquired strengths, talents, gifts, habits, practices, and institutions—were positive. They made it possible for society to fight disease, limit population, increase production, enlarge individual and social capacities—in short, to create civilization. Granted, this was not what Lamarck had in mind in his *Histoire naturelle* or in his *Philosophie zoologique*. But then neither was social Darwinism what Darwin had in mind in his *Origin of Species* or *The Descent of Man*.

Technology is the most effective, as it is the most familiar, example of an instrument of Lamarckian evolution. It is effective, not because it relies passively on the survival of the fittest or on the inheritance of acquired physical characteristics, but because it depends instead on conscious imitation, adaptation, expansion, experimentation, and education, and it does not develop gradually, through tens of thousands of years, but grows exponentially. Where Darwinian theory encouraged resignation or passivity—after all, the processes of nature could not be hurried—Malthusian theory conjured up despair, and both taught the impropriety of government interventon, technology calls insistently for government cooperation and, in the interests of society, intervention.

Like science, technology is neutral. It is man who applies it, and it is man who is responsible for it—man working alone, or through corporations or government. Across the long arch of time, from the first century B.C., when Pollio Vitruvius found "traces on the Rhodean Shores" and wrote that it was the inventive and psychic qualities in man that accounted for civilization, to Hiroshima, technology has been the chief instrument for progress, or for death, as man has chosen to use it.

We have no choice but to control our technology; if we do not, it will control us. The history of the last half century makes clear that technology, like science, is ecumenical; its character and conduct cannot be defined by any one nation, only by a concord of nations. This was implicit in Churchill's famous verdict on the atomic weapon: "We are now placed in a situation both measureless and laden with doom, for now safety will be the sturdy child of terror, and survival the twin brother of annihilation."

I close with some lines from the most venerable of our philosophers. In what may have been his last public address, when the great controversy was just getting under way, Ralph Waldo Emerson wrote:

> Power can be generous. The very grandeur of the means which offer themselves to us should suggest grandeur in the direction of our expenditure. If our mechanic arts are unsurpassed in usefulness, if we have taught the river to make shoes and nails and carpets, and the bolt of heaven to write out letters, let these wonders work for honest humanity, for the poor, for justice, genius and the public good. Let us realize that this country, the last found, is the great charity of God to the human race. America should affirm and establish that in no instance shall all the gains go in advance of the present right. We shall not make *coups d'état* and afterwards explain and pay ... but shall proceed like William Penn, who treats with the Indians, or with the foreigner on principles of mutual advantage; the Constitution ... should be mankind's Bill of Rights, or Royal Proclamation, announcing that now, once for all, the world shall be governed by common sense and love of morals.

Privacy and Autonomy
in American Law

A. E. DICK HOWARD

L egal protections of privacy mirror the social values of a society. Cultural anthropologists have noted that the Germans—a people given to walking in their highly valued forests—tend to claim a larger sphere of privacy than do many other Western peoples. This claim is reflected in the German law that prohibits photographing strangers in public without their consent. The casual traveler abroad may observe different attitudes toward privacy from one country to another; there is, for example, the famous reserve that many Englishmen display in their railway carriages and other public places.

A look at American social and cultural history reveals conflicting imperatives. On the one hand, no people on earth are more fond than Americans are of proclaiming their individuality; it was not the English or the Germans who produced the Marlboro Man. America's enormous territory and the availability, especially during the nineteenth century, of new frontiers to which to migrate, encouraged claims of one's own personal space. Yet, especially in puritan New England, there was a strong social tradition of dictation and supervision of private behavior. Sexual conduct, in particular, was a matter of common concern, not of individual choice.

A right to privacy was not part of the English common law, nor was it specifically recognized in early American law. The idea of legal protection for claims to privacy received its most significant impetus with the publication, in 1890, of an article by two young lawyers, Samuel D. Warren and Louis D. Brandeis, the latter eventually to become a justice of the United States Supreme Court. Noting the advent of yellow journalism and the appetite of the mass-circulation newspapers for scandal, Warren and Brandeis said, "Gossip is no longer the resource of the idle and the vicious, but has become a trade, which is pursued with industry as well as effrontery." They urged, therefore, that citizens be able to recover damages for unreasonable publicity. In the years since 1890, courts in most states have adopted the principle of a right to privacy. Thus damages may be recovered by someone whose name or picture is used for commercial purposes without his consent, or who is placed in a false light by publicity. A person's seclusion may be protected; a famous modern case is that of the protection given by a court to Jacqueline Onassis from the harassment of an overeager free-lance photographer, who regularly lay in wait to take pictures of her and her children.

Few concepts in American law have proved to be so malleable as that of privacy. It is capable of no simple definition. Instead, it has taken on a number of forms. Perhaps the most obvious claim to privacy has to do with controlling what others know about a person. Alan Westin, of Columbia University, has defined privacy as "the claim of individuals, groups, or institutions to determine for themselves when, how, and to what extent information about them is to be communicated to others." Although it nowhere uses the word *privacy,* the Constitution protected some aspects of this kind of privacy a century before the article by Warren and Brandeis. A chief complaint of American colonists against British policy was use by the authorities of writs of assistance—general warrants used to search for evidence of customs violations. Hence the Fourth Amendment to the U.S. Constitution, which forbids unreasonable searches and seizures. Another constitutional protection for information privacy is found in the Fifth Amendment guarantee of the privilege against self-incrimination.

Sometimes an individual resists intrusions, not in order to restrict access to information about himself, but simply in the interest of being left alone. He may want to have peace and quiet, hence may object to sound trucks in his residential neighborhood. Or he may resist being an unwilling audience for unwanted messages or images, such as music and advertisements piped into buses. He may dislike being the unwilling recipient of obscene materials sent through the mails, or he may prefer not to have door-to-door solicitors disturb his repose. All these claims have come to the surface in court decisions, many of them requiring judges to balance the interest in privacy against claims that the First Amendment protects the right of another party to convey his message even to those who may not want to hear it.

Claims of privacy in modern American law have outgrown these more conventional examples that turn upon the wish to keep information to oneself or to be left alone. Changing American mores—especially in regard to choices about what has come to be called lifestyle—have added larger dimensions to notions of personal choice. Far from being essentially passive claims, these involve choices about personal behavior. Some, such as decisions about sexual conduct, may take place in private circumstances, but others, such as decisions about personal appearance, do not. In either instance, there is a claim to do as one pleases, free from interference by the state.

The debate over personal autonomy—over what the free spirits of the 1960s called "doing your own thing"—did not originate on

American college campuses. In his classic essay *On Liberty* (1859), John Stuart Mill asked whether there was a sphere of personal autonomy that the state and the law should respect: "What, then, is the rightful limit to the sovereignty of the individual over himself? Where does the authority of society begin?" Mill's answer: "The sole end for which power can be rightfully exercised over any member of a civilized community, against his will, is to prevent harm to others."

Efforts to craft legal protection for notions of personal autonomy often take the form of constitutional arguments. The Constitution says nothing about the right to use contraceptives, the right to marry whom one pleases, the right of consenting adults to have sex in the way they prefer, or the right to wear one's hair any way one likes. So petitioners have looked to the more general language of such constitutional provisions as that which accords liberty the protection of due process of law.

Contemporary uses of the due-process clause commonly arise in the context of individual behavior having nothing to do with economic enterprise. Precedents for using that clause to protect rights not spelled out in the Constitution, however, existed in the uses made of it by conservative lawyers and judges in the late nineteenth and early twentieth centuries to defend economic enterprise and laissez faire against state interference. A famous example is the 1905 decision of the Supreme Court in *Lochner v. New York*, which struck down minimum wage and maximum hours legislation as infringing "liberty of contract."

Sixty years later, the Supreme Court opened a new era in personal autonomy cases. In *Griswold v. Connecticut* (1965), the Court overturned the conviction of defendants, including a doctor, who had been charged under Connecticut law with giving information and medical advice to married persons on means of preventing conception. Justice William O. Douglas, announcing the decision of the Court, found a "zone of privacy" formed by "emanations" from explicit guarantees in the Bill of Rights. Justice Hugo L. Black, dissenting, thought the Connecticut law "every bit as offensive" as did his brethren, but he did not think that made it unconstitutional. "The Court talks," he said, "about a constitutional 'right of privacy' as though there is some constitutional provision forbidding any law ever to be passed which might abridge the 'privacy' of individuals. But there is not."

Griswold quickly became a lodestar for litigants hoping to find constitutional protection for other kinds of behavior. The concept of a protected zone of privacy received furher explication in 1969 in *Stanley v. Georgia*. State and federal agents, looking for evidence of bookmaking in Robert Eli Stanley's home, had instead found three reels of pornographic movie film. Stanley was convicted for knowing possession of obscene matter, and the state defended the conviction by arguing, in effect, "if the State can protect a citizen's body, may it not also protect his mind?" The Supreme Court, reversing the conviction, responded that "a State has no business telling a man, sitting alone in his own house, what books he may read or what films he may watch. Our whole constitutional heritage rebels at the thought of giving government the power to control men's minds." The justices may well have been thinking of George Orwell's *Nineteen Eighty-Four.*

After 1969 the composition of the Court changed significantly. By 1972 President Nixon had appointed four justices, one of whom, Warren Burger, replaced Earl Warren as Chief Justice. It might have been supposed that a tribunal in many ways more conservative than the Warren Court might be reluctant to expand the zone of personal autonomy and privacy further. In particular, "conservative" justices might be slow to use judicial power to discover rights for which, as Black had pointed out in *Griswold,* the Constitution offered no explicit textual support—especially when the cases involved behavior that might offend traditional notions of morality.

It is striking, therefore, that in 1973 the Court, in *Roe v. Wade,* held that the due-process clause of the Fourteenth Amendment protects a woman's right—a privacy right—to decide whether to have an abortion. The scope of the privacy right in *Roe* goes far beyond that declared in *Griswold.* The 1965 case was more nearly concerned with privacy in the traditional sense, involving the intimacy of the marital bedroom. The claim in *Roe* was manifestly one of personal autonomy—the right to make and carry out the abortion decision without state interference. *Griswold* and *Roe,* moreover, differ in the nature of the interest the state sought to invoke. In *Griswold,* the state was hard pressed to show that a persuasive interest was served by regulating the contraceptive practices of married couples. In *Roe,* by contrast, the state could claim that it was protecting an incipient life, that of the fetus.

Protection of personal autonomy extends beyond such intimate decisions as contraception and abortion to behavior that takes place in quite public situations. In a 1972 decision, *Papachristou v. City of Jacksonville,* the Court invalidated a local ordinance under which, as Justice Douglas said, "poor people, nonconformists, dissenters, idlers" might be required to comport themselves "according to the lifestyle deemed appropriate by the Jacksonville police and the courts." In *Wisconsin v. Yoder* (1972), the Court vindicated the preferred lifestyle of the Amish by upholding their challenge to Wisconsin's compulsory attendance law.

Even modern modes in speech—including expressions that some

find offensive or tasteless—have received constitutional protection. When Paul Robert Cohen, an opponent of the Vietnam War, entered the Los Angeles courthouse wearing a jacket bearing the words "Fuck the Draft," he was arrested and charged with breach of the peace. Reversing Cohen's conviction, Justice John Marshall Harlan observed, in *Cohen v. California* (1971), that "one man's vulgarity is another's lyric." In another case, *Eaton v. City of Tulsa* (1974), Justice Lewis F. Powell, Jr., commented, "Language likely to offend the sensibility of some listeners is now fairly commonplace in many social gatherings as well in public performances."

Legal protection for privacy and autonomy bolsters a number of values. It fosters individualism and human dignity by emphasizing the uniqueness and worth of each person. It reinforces intimate relationships, especially marriage and family. Protection of privacy or autonomy is especially important to racial, religious, or other minorities whose beliefs and practices may be distasteful to a majority. It encourages an open society, especially in the political arena, by protecting dissent and unorthodoxy. In so doing, legal sanction of privacy and autonomy operate in the tradition of the American constitutional system itself, one based on Lockean assumptions about natural rights and on Madisonian precepts about limited government. Consideration of the libertarian dimensions of the American tradition makes it easy to see why the idea of privacy and autonomy has been applied in so many legal and judicial contexts.

In law as in life, however, there are few absolutes—death, taxes, and the fortunes of certain baseball teams aside. Claims of the right to privacy or autonomy often collide with other legitimate social interests. In particular, the personal autonomy cases raise the issue of the extent to which the law may be used to enforce the judgments of society about morality. A distinguished British jurist, Lord Devlin, has argued, in *The Enforcement of Morals* (1965), that society is entitled to use law to enforce its "common morality"—Devlin's answer to those who would "decriminalize" such crimes as homosexuality and prostitution. It is in the spirit of Devlin's analysis that the Supreme Court, in *Paris Adult Theatre I v. Slaton* (1973), upheld the authority of a state to regulate the exhibition of obscene materials in "adult" theaters. Rejecting the argument that only consenting adults were involved—and that therefore the reasoning of *Stanley v. Georgia* should apply—Chief Justice Burger emphasized the interest of the public in the "quality of life" and its right to "maintain a decent society."

Thus any effort to generalize about the legal protection of privacy and autonomy—including that sought to be drawn from the Constitution—has to be strongly qualified when legislators and courts recognize some competing social interest. Even when the Supreme

Court acknowledges the privacy or autonomy right, that right may be limited by some competing interest. Thus, while the woman's right to decide to have an abortion, as described in *Roe v. Wade,* is absolute in the first trimester of pregnancy, the Court ruled that a state may look to the interest in maternal health to regulate abortion procedures in the second trimester and may invoke an interest in impending life to curtail abortions in the third trimester—that is, after the point of fetal viability—even more sharply.

Other social interests may qualify or even deny a person's claim to privacy or personal autonomy. Adultery laws undoubtedly limit a person's choice of sexual partners, but there is little doubt of their constitutionality, in view of the interest in marriage and the family as basic social institutions. Obscenity laws, while they must conform to the requirements of the First Amendment, in order not to impinge upon protected speech, may be upheld without the state's being obliged to prove that those who read obscene materials are more likely to commit antisocial acts. The state may often even be paternalistic; the vast majority of courts that have passed upon the constitutionality of laws requiring operators of motorcycles to wear helmets have upheld those laws, rejecting the argument that the state has no business protecting the individual from his own folly. Sometimes a claim of autonomy will collide with another person's rights of privacy; thus my wish to express my individual personality by shouting four-letter words—or singing arias from Italian opera—in my neighbor's window may founder upon his wish for peace and quiet.

Developments in technology have further complicated efforts to give legal definition to privacy and autonomy. Modern technology did not exist when British authorities used writs of assistance in the American colonies; the techniques of electronic surveillance require today's judges to decide whether the Fourth Amendment protections against search and seizure apply to methods totally unknown to the framers of the Constitution. In *Katz v. United States* (1967), the Supreme Court held that electronic wiretaps fall within the ambit of the Fourth Amendment. In *United States v. White* (1971), however, the Court held that neither a warrant nor probable cause is necessary for third-party electronic eavesdropping when one party to a conversation—usually an agent wearing the device—consents. And in *United States v. Knotts* (1983), the Court found no Fourth Amendment violation in the monitoring by the police of beeper signals emitted from a device attached—without the owner's knowledge or consent—to a private automobile.

These Fourth Amendment cases are but one example of the way the effects of technology must be taken into account in considerations of privacy and autonomy. The expansion of data banks, of information technology, of surveillance techniques may greatly alter the way

A. E. Dick Howard

individuals look at conventional transactions. The ordinary events of daily life—automobile travel, telephone calls, check cashing, even social conversations—are not so ordinary when the possibilities of technology are taken into account. Transactions generally supposed to be casual and isolated may no longer be unrecorded or forgotten. What might be expected to be known only to another party to a transaction may prove to have a larger audience, remote in time and place. Memories fade, but information retained by technology has permanence.

As technology enlarges the scale on which actions, words, or other transactions may be known and recorded, individual expectations of privacy may shrink accordingly. This may have unsettling implications for a person's perception of himself as autonomous. To the extent that free government, especially in the American constitutional system, turns on notions of individualism and civil liberties, to that extent the ability of citizens to see themselves as having a zone of privacy bears directly upon the health of the body politic.

Privacy and autonomy are, of course, not the only values in a free society. Others must be served as well, and they may conflict with or ultimately qualify claims to privacy or autonomy. But having such a zone of privacy attached to the individual sustains his opportunities for reflection, self-expression, and appreciation of his own worth. In Orwell's *Nineteen Eighty-Four,* "You had to live—did live, from habit that became instinct—in the assumption that every sound you made was overheard, and, except in darkness, every movement scrutinized." Present-day technology, permitting seeing in darkness, would eliminate even the exception. But, exceptions or not, technology does not stand still, and its availability—even for benign and useful purposes—requires those who care about human worth to understand the relation of privacy and autonomy to a free society.

Education for Utopia

O. B. HARDISON, JR.

I had a friend who was a professor of pyschology at a large university. He would always begin his classes by accouncing: "Students, I want to be honest with you. Therefore you should know that half of everything I will tell you in this course is lies. My problem is that I do not know which half."

To talk about education, one must talk about education for something. And to talk about the something, one must make assumptions about the future. Fifty years ago it was possible to define the something with considerable assurance. Things changed slowly—by generations rather than decades. Barring a war, the future would be more or less like the past.

Today, things are changing rapidly, and the pace of change is accelerating. Anyone who talks about education is therefore compelled to be a prophet, and prophecy is a risky business. No matter how probable the forecast, it will be half true if it is inspired and one tenth true if merely mortal.

Let us call the future toward which we are inexorably moving Utopia—not because we are certain that it will be a desirable future, but because Utopia means "nowhere" in Greek. We cannot know what the future is until we have reached it. We only know that we are in a state of rapid motion. Things change from decade to decade, even from year to year.

The changes are so rapid, so fundamental, and so numerous that they can be used to justify any number of predictions. Consider the deep and unpredictable changes caused by what once seemed a relatively simple innovation, the birth-control pill. Margaret Sanger thought that birth control would strengthen the family. Perhaps it has. But it has also permitted a sexual revolution and thus has changed attitudes toward the family that are thousands of years old. As attitudes have changed, the position of women has changed generally, I think, for the better. The position of children has also changed, both for the better and for the worse.

Birth control has also had countless other effects. By reducing the birth rate in the developed countries it has helped to change their demographic profiles. This effect has been reinforced by improvements in health care that have extended the average human life span in the developed countries by 25 percent since 1900. Demographic change, in turn, affects all social planning—social security and pensions, for example. It affects politics also. The elderly vote in larger numbers than the young, and there are more of them.

Society is a seamless web. To introduce methods of birth control is not simply to clear away an obstacle on the road to the good life. It is to change the definition of the good life and, simultaneously, every other relation in society.

Today the scientific world is a ferment of technologies that promise even more revolutionary social consequences. The most dramatic, genetic engineering, is a fact of modern life, although its effects are probably a decade away. It promises to make organic life as malleable as inorganic matter—with consequences that no one can imagine.

Robotics, another technology, promises to revolutionize manufacturing. It is already under way, and I predict that it will move forward at a pace that will astonish even its most enthusiastic practitioners. A recent analysis of the robotics industry by Prudential-Bache notes that there were 1,450 robots on line in the United States in 1980. It suggests that by 1990 there will be more than 20,000. I believe that the number in 1990 will be more than 100,000, and that they will be doing jobs that are now only a gleam in the eye of the robotics engineer.

Robotics is a special case of the information revolution. This revolution is now changing American life more rapidly than any technology that has ever appeared in our history. It is rapidly climinating routine jobs and it promises similar, though more gradual, reductions in the need for highly trained professionals.

Whatever Utopia we are approaching, the middle passage between here and there may, many believe, produce enormous dislocations of the work force. In "The Declining Middle," a much-discussed article in the July 1983 *Atlantic Monthly,* Bob Kuttner predicts a society divided between a majority of poorly paid clerical workers and a small technological and managerial elite. "An industrial society," he writes, "employs large numbers of relatively well paid production workers. A service economy, however, employs legions of keypunchers, sales clerks, waiters, secretaries, and cashiers, and the wages for these jobs tend to be relatively low." Kuttner's projection is echoed by a report, "The Future of Work," issued in August 1983 by the Executive Council of the AFL-CIO. The report argues, among other things, that

> as computers and robots take over more and more functions in the factory and office, a two-tier work force is developing. At the top will be a few executives, scientists and engineers, professionals and managers, performing high-level, creative, high-paid full-time jobs in a good work environment. At the bottom will be low-paid workers performing relatively simple, dull, routine, high-turnover jobs in a poor work environment.

This sounds a lot like *Nineteen Eighty-Four*. Is not the peonization of the American work force already in progress? What other interpretation can be given to pay-backs by Ford and GM automotive workers? To "intimidation by bankruptcy" by Chrysler, Continental Airlines, and, most recently, Eastern Airlines? To union-breaking by the FAA and Greyhound?

Consider another ominous fact. In a survey of 20,000 graduates of the high-school class of 1972, Thomas J. Moore found that 43 percent of those who attended college are employed in jobs that do not require a college education. He adds that "there is every reason to believe that the situation in 1983 is worse than 1979" and that it will get worse rather than better.

Kuttner, Moore, and the AFL-CIO believe they are pessimists. I believe they are optimists. It is clear that production jobs will decline during the next decade, as people are replaced by machines. It is by no means clear, however, that "simple, dull, routine" service jobs are safe from automation. Optical scanning and source-input are already making keypunching anachronistic. Sales clerks and cashiers are being replaced by credit card machines, bar coding, automatic tellers, and direct mail catalogues, all of which are now graduating from the printed page into electronic marketing networks.

Secretaries, as well as file clerks, bookkeepers, purchasing agents, payroll clerks, special librarians, and the like, are being replaced by microcomputers that can read printed and typewritten text and are beginning to be able to read freehand printing and recognize voice commands. CAD-CAM—that is, computer-assisted architectural and engineering design—is replacing draftsmen. Because of cultural lag, there will continue to be numerous jobs in all these areas, and their number may even increase for a few years. Eventually, however, the curve will flatten and tip downward.

The point is that the next two decades will see an inexorable reduction in the number of "simple, dull, and routine" jobs that have absorbed large numbers of workers in the past. Precisely because many low-level jobs *are* repetitive and require little creative intelligence, they are prime targets for automation. And as soon as one large firm in a given industry replaces its clerical workers with machines, all other firms in the industry will be forced to follow suit or become noncompetitive. It is difficult to imagine how unions will be able to counter this trend, since their only recourse will be protectionism. But protectionism in an interdependent world economy is an exercise in futility.

But why stop with the underclass? Artificial intelligence is changing the conditions of professional work through so-called expert systems. Expert systems use the state-of-the-art knowledge of a given subject to increase the productivity of experts. Expert systems are now on line for jobs as diverse as analyzing cloud-chamber images of the paths of subatomic particles, deriving molecular structures from radiographs, analyzing geological information for prospecting, diagnosing medical problems, finding precedents for legal briefs, and maintaining plants and machines.

Expert systems will reduce the demand for human experts in the developed countries and will close the gap in specialized knowledge that now exists between the developed and the underdeveloped countries. They will permit many problems that now must be handled by specialists to be handled by paraprofessionals.

We are evidently approaching a period during which wealth will increase, while the number of jobs in traditional high-employment areas will decline, no matter what government policy is followed. This will be a worldwide phenomenon. The most immediate needs of society will be met by a small number of workers supported by automation and information technology.

It will be a permanent social fact. The AFL-CIO report, "The Future of Work," is not very different, really, from Orwell's vision of a society divided between a party elite and a vast, leaderless mass of proles. If his projection is valid, there is very little to be said about the future of education. A minority of citizens will be educated as experts. The rest will have to be taught to be docile.

But is the projection of "The Future of Work" persuasive? Two historical parallels are useful here. At the beginning of the twentieth century, approximately half the American labor force was agricultural. Today about 5 percent of the labor force is in agriculture. Such a massive change in work patterns could have been violent and bloody. Instead, it was relatively peaceful, although it caused widespread suffering.

More recently there has been an equally dramatic shift in the urban labor force from production to service. The Census Bureau now estimates that only 25 percent of American jobs are in production; most of the rest are in service occupations. Again, the transition has not been without suffering, but it has been remarkably peaceful.

I do not predict the disappearance of traditional jobs, much less the immediate disappearance of such jobs. I do predict a leveling off of demand in a great many areas that have traditionally provided large numbers of jobs, followed by a gradual, then a more abrupt, decline in such jobs. The process will take at least two decades, and it will be erratic. It will be sudden and extreme in some areas but almost imperceptible in others.

The first lesson to be drawn from the historical shifts in work patterns is that society can accommodate a remarkable amount of change without disintegrating. A second lesson is that with a little understanding, a great deal of suffering can be avoided. A third lesson is that as labor is released from one area, it tends to be absorbed in another. Consider each area of the economy as a project. In general, we move from projects that are necessary for survival to projects that are desirable to improve the quality of life. It is as though society were saying, "Since we can now meet the needs of survival, let us devote the workers who have been released from survival jobs to projects that make survival worth while." To ask how workers released from service jobs will be employed is to ask what the next great human project will be. It is not to ask how unions can shelter jobs that are threatened by automation and foreign competition, but ask how unions can shift from their historical function as production-specific lobbies to a potential function as partners in the effort to plan America's future. It is also to ask what people can do better than machines.

Machines can work longer hours, in more arduous surroundings, than people can. They are not distracted when performing repetitive exacting tasks. They can remember more information more accurately than people can, and they can manipulate this information more rapidly. It is common to say that they are stupid, but they are rapidly getting smarter.

But people are superior to machines in relating to other people, and perhaps the next great growth area in employment will be in the interpersonal sector. The interpersonal sector is not new. It is as old as work. It is the sector of the economy in which people are paid to interact with other people. It consists of all those jobs that require caring, instructing, advising, mediating, managing, and creating.

Such jobs are now generally classified as service jobs, but, unlike most jobs in the service sector, they are not routine and mechanical. Machines cannot advise people about their personal problems, mediate disputes, provide counsel when the mediation fails, relate unique motives and feelings to general objectives, negotiate differences between nations, or help the young to relate information to values and creative goals. Only people can.

Society has always had counselors, lawyers, therapists, rehabilitation experts, politicians, diplomats, teachers, and artists. It has never, however, had enough of them, and ready access to their services has traditionally been limited to the wealthy. Society has tried to remedy its interpersonal problems through charity and social welfare, but its success has been limited. Even in the wealthiest countries, children are inadequately educated, adults become unproductive and antisocial because of physical or psychological trauma, injustices generate hostilities between individuals and classes, and creative energies that could benefit society are turned inward.

I think that there is an almost limitless potential for growth in interpersonal work, and I believe that the interpersonal field is the one most likely to absorb workers displaced by machines from the production and service economies.

Let me place this idea in historical context. Underlying most of the political ferment of the 1960s was a passionate longing for a just society, a society that lived up to the ideals of the Constitution and Bill of Rights. It was sudden and violent, and because of its suddenness and violence it was self-defeating. But whatever its failings, it forced Americans to face up to the enormous agenda of their unmet needs.

Perhaps the interpersonal sector will meet these needs. Perhaps the next search for a just society will be long on accomplishment, and—because we may have learned something the first time around—short on the kind of rhetoric that polarizes society. Perhaps we can create a society that combines material affluence and access to knowledge with the human dimension that has been in short supply in the developed countries since the industrial revolution.

What sort of an education fits this image of Utopia? Consider the most common explanations of what education is supposed to do. It is supposed to prepare people for jobs. It is supposed to give them a sense of their relation to society. And it is supposed to develop their understanding and spiritual resources.

I have no quarrel with education for jobs if they are the right jobs. I am disturbed, however, by the hard-edge practicality of current discussions of education. At present, the two hottest educational topics are "basics" and computers. There is nothing wrong with basics if basics means the tools that are the necessary prelude to education. But if education stops at "reading, writing, and computational skill," it is simply not of any real value or significance for the late twentieth century. Education in an advanced, rapidly changing technological society *must* be concentrated on the development of understanding of the way skills are used creatively once they are acquired.

The current infatuation with computers is even shallower than the basics movement. By the time the grade schoolers who are now tapping away on Apples and Commodores graduate from high school, their alleged computer skills will be obsolete. This is also true, unfortunately, of a great many of the narrowly vocational programs that students are being encouraged to pursue, whether in high school or community college or graduate school.

If the interpersonal sector is the next growth area, education for jobs must stress the development of interpersonal skills. This means that in addition to being informed, students will need to develop self-confidence and sensitivity to the needs of others. These qualities can only be taught—*nurtured* would be a better word—by teachers who regard students as individuals, are concerned about their spiritual growth, and are themselves skilled in interpersonal relations. And this kind of development must take place in classes that are small enough for students to receive regular individual attention.

Since personal development is inseparable from development of a sense of values, education for the future should be concerned with values. I do not mean that it should be ideological or that it should be used to indoctrinate. Quite the opposite. It should engage the student in a critical examination of values as they have been expressed in history, in art and literature, and in science, philosophy, and political theory. It should emphasize the specific heritage of the society in which it is taking place. In America education should emphasize the Western tradition that begins in Palestine and Greece, but the place of the West in world society should also be considered. Its purpose should be to encourage inner growth, which requires both an affirmation of the values that have shaped the student's identity and a liberation from the narrow and parochial aspects of that identity.

Perhaps the most interesting kind of education for the future is education for the fun of it. Somehow, in the grim debate on basics, it is seldom mentioned that education at its best is exhilarating and that the excitement often lasts a lifetime. It is the same sort of excitement that is associated with a challenging game, even a computer game. It is experienced by carpenters, birders, flower gardeners—What, after all, could be more useless than a bed of petunias?—professors of medieval literature, concert violinists, physicists, and model railroaders. It is the secret of the most successful kind of education, and without it education becomes a dreary transfer of information from one bucket into another.

An educational system that is concerned about the spiritual growth of each student, that teaches values as well as facts, and that knows knowledge to be enjoyable as well as useful is the best sort of system that I can imagine for the future we seem to be approaching.

Red Tape Unraveling

Eliot D. Chapple

"We cannot think first and act afterwards. From the moment of birth we are immersed in action, and can only fitfully guide it by taking thought."
—Alfred North Whitehead

"Hold your tongue!" said the Queen, turning purple, "I won't!" said Alice. "Off with her head!" the Queen shouted at the top of her voice. Nobody moved.
"Who cares for you?" said Alice (she had grown to her full size by this time). "You're nothing but a pack of cards!"
At this the whole pack rose up into the air, and came flying down upon her; she gave a little scream ... her sister ... gently brushing away some dead leaves that had fluttered down upon her face."
—Lewis Carroll
Alice in Wonderland

Today, we are told, we are in the early stages of a new era in our society when the telecommunications media will assume supreme command. But the significance of high technology is overlooked if our understanding of it is limited to two-way television and CB radios, or to word processors plus computer memories to aid our spelling. However they are processed or transmitted, it is words and ideas that are crucial. We can manage communication far more efficiently and expeditiously, but what we say, at whatever level, persists in ambiguity. Microchips, lasers, and fiber optics are part of the technological elaborations of today, but the raw input is still human speech and human idiosyncrasy. We can radically modify the space-time constraints of the channels that link persons together. Indeed, this is a significant component of the high-technology revolution. But rarely do we recognize how science and technology are thus used in shaping the interactions of individuals.

What I think all true scientists have in common is a bias toward observation or, in archival material, toward what might be established as an observable. We want to see what people actually do, by themselves and with others, in different situations and environments. It is this which increases our confidence in the reliability of cross-cultural and cross-historical studies. We try to identify and understand the patterns of interaction that restrict or facilitate that individual's freedom to act, to initiate freely, to be his own boss. Since we begin with daily life—realistically, we end there too—we have to be continually concerned with estimating the effects of the environment, physical or man-made, technology, and the cultural patterning of activities on the interactions of individual personalities. Changes in any of these components bring about changes in the organizational system that constitutes the cultural milieu or matrix within which human life and human potentialities have to be realized.

But what is fundamental if we are to understand the motive forces behind such changes is the realization that the rhythmic patterns of action and interaction of individuals have primacy. They are the continuing results of the transformation of the energies constantly produced by the body and mobilized for action by the central nervous system through its somatic, autonomic, and endocrine subdivisions. It is these properties of biological organisms that establish the dynamics for exact scientific formulations of the process of developing and elaborating relations to others and to the environment. Analysis is facilitated remarkably by the fact that what on the surface appears to be an irregular rhythmic occurrence of actions and interactions in so much human activity is, in fact, built out of the complex interplay and integration of populations of biological oscillators. Today we speak of circadian clocks, but in fact, body clocks are fast and slow, at every level of biochemical and physiological functioning. They combine into total subsystem reactions that we call the emotions, crudely categorized, without ever asking how the whole organism is energized and why.

We need to define the conditions under which individuals can enjoy a heightened and continuing state of freedom. Though I am sure we all agree, with Orwell, that this is what most of us hope to achieve, we must still establish objective criteria for determining whether the cultural patterns available in any society provide the means to the ends we are seeking. In tightly controlled and rigid systems, people's relations are narrowly circumscribed. Freedom to initiate, and more than that, to interact with others for even brief periods outside a closely specific network, threatens the higher-ups with an uncontrollable development of competing systems and a potential decline in the frequency and even the certainty of automatic responses to the initiatives from on high.

Given the remarkably idiosyncratic divergences among human personalities, men as animals are inevitably forced—against their will, we say—to synchronize their behavior to the limited range of inflexible patterns required by totalitarian societies. But history shows that even the true believers find that the inner dynamics of network struggles changes their potential participation in the shifting bureaucratic system. To maintain a position as dictator requires selecting certain types of personality over others, as environment and technology go through their own, presumably unforeseeable, changes. Since human beings are only peripherally concerned with reason, it is the explosive force of biological determinism in reacting for or against others emotionally-interactionally that controls behavior. Only where the cultural environment reduces or minimizes the barriers to free initiation, in accord with the needs of one's own rhythms, can that distant drummer be heard providing the beat for a new commitment.

If we take a hard look at present and past societies and ask ourselves why it is that high technology, if extended across the board, might well bring about significant changes in the way we live, work, and will be governed, we must realize that, in itself, high technology won't do our work for us. There are many today who seem to believe that the consequences of technology will be to provide an aery blanket on which we can float through our lives in blissful peace. Others think that it will only set the stage for a greater tyranny than the world has ever known. But it is difficult to see how totalitarian societies will be able to accept the changes necessary to produce the benefits of high technology; maintaining political control requires increasing, not decreasing, centralization and enforcing compliance to a rigid protocol.

High technologies can facilitate the growth of human freedom only because they can radically reduce the asynchronies in our relations with others and the stresses these asynchronies build up. They do so because they make it unnecessary and ineffective to continue relying on erratic channels of communication and transportation; by eliminating the endless cultural steps that paper-flow compilation requires, we enable smaller groups of people to interact more intensively and harmoniously in highly simplified systems of organization. This becomes possible, however, only if the organization that uses high technology has all the materials, the techniques, and the cultural inputs on the site. By reducing spatial diversity and extreme division of labor, hierarchical, or bureaucratic, centralized controls become redundant.

For centuries, nations have fought to coordinate and centralize their power over all those within their boundaries. Inevitably, they see stability as threatened, implicitly or explicitly, by those outsiders with whom their own people interact. Thus they try to strengthen themselves by territorial expansion or manipulation of trade. And this requires bureaucracy. Its driving force, centralization, has been accelerated in recent years by the imposition of the paper revolution on the already awkward organizational consequences of the industrial revolution. Just as mass production, once engineered, has made organizational copies possible throughout the world, so engineering the paper depends on treating the individuals to whom it is directed as stereotyped replicas of a minute segment of their cultural contexts. Laws, rules, regulations, procedures are devised for the largest number of persons who can be given some cultural classification. Paper, therefore, is produced to establish least common denominators among the citizenry and thereby to facilitate controls over those within large organizational networks. These are justified by elaborating the logics of financial custom and legislative and regulatory fiats. The exceptions—and all of us in one way or another can become exceptions— are avoided or neglected, unless countervailing pressures can be brought to bear, enabling human beings to assert their individualities and the biological rhythms on which these are built.

So to transform traditional societies, we need to cut the strings by which the central power or establishment imposes uniformity— or at least the appearance of uniformity—in accepting directions from on high. The appeal of ritualized subordination is reinforced as supposedly invariant habit by inducing or compelling every person to take part in great ceremonies under the careful guidance of the leader. So hierarchs share to lesser and lesser degrees his charisma, depending on their relative positions in the network from the top. But the rituals do not override individuality. They have strength only if the rhythms of personality can be harmonically bound to the integrals of the ceremony.

Human animals are not driven genetically to cohere in such mass herding. The biological properties of the species limit the number of persons who can interact effectively with one another and the amount of sensory information they can receive and react to. Unlike the herring, swimming in schools, shifting direction and velocity but never losing formation, we are incapable of mass existence. The suddenly continuing crowd, a consequence of prevailing and cumulating disharmonies and cultural dislocations, has only a temporary and intermittent life, however ghastly the violence liberated in its brief and frenzied state.

Far more today than ever before, mobilization of monies for international use is creating a highly integrated, worldwide system, subordinating the productive efforts of individual nations to it. Technology and the paper assumed essential to hold it together are selected to strengthen control by the leaders and accelerate centralization. In the process of readjustment to change, takeover by crowds has to be followed by stabilizing groups to pick up the pieces, however briefly.

Too often, acquisitions asked for are incongruent to the needs, but the demand for funds escalates. In consequence of this intensified competition for monies to spend, the lenders who, by necessity, manage the paper are being forced to ask the keepers of these rituals—accountants and economists—to shift the assumptions to make it possible to separate the individualized organizational systems for differential treatment. It has become shockingly obvious that large loans to confused political and governmental units, and to private units as well, are highly vulnerable to the intractability of human beings. They never follow the rational rules of abstract economic logic, which the cultural standards of American and European businesses assume, though they rarely follow them themselves.

Even if the evidence may still seem sporadic to communication specialists imbued with the Protestant ethic, modern managers are now coming to realize that recognizing the biological need of every person for freedom of initiative is the primary requirement for the achievement of high productivity. How long it will take for paper engineering to undergo this necessary revision on the basis of the individual as a coherent system is anybody's guess. Yet those whose predilection is for observing their fellows are quite aware that effective managers are recognizing this as essential, even if they are often uncertain what steps they need to take to make it reality.

DILEMMAS

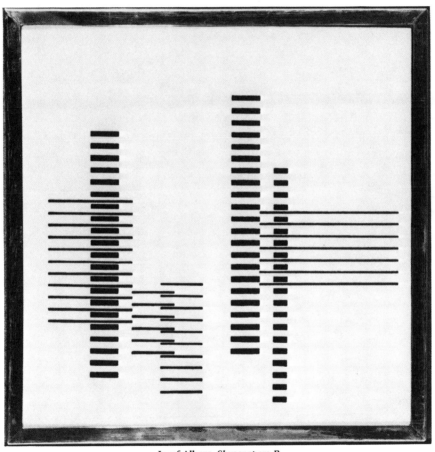

Josef Albers, *Skyscrapers B*

The Dilemma of Civic Learning in the Schools of a Pluralist Democracy

R. Freeman Butts

"It is an axiom in my mind that our liberty can never be safe but in the hands of the people themselves, and that too of the people with a certain degree of instruction. This it is the business of the state to effect, and on a general plan." —Thomas Jefferson

The United States has been inundated by a rising tide of public discussion about the plight of American education. A succession of reports by high-level national commissions and task forces, diagnosing what ails the schools and prescribing treatment for their improvement, has touched a remarkably responsive chord across a wide spectrum of public opinion. These reports have loosed a flood of articles and editorials in the press, commentaries on the talk shows of TV opinion molders, and debates in legislative halls as well as in academic and professional circles.

The password for educational reform is most often summed up in the word *excellence,* a word that few in the public or in the profession would be able to define or disagree on. It is most often described in fairly concrete, simplistic ways that the public can easily understand:

- increasing the number of courses required for graduation from high school, especially in science, mathematics, and computers;

- prescribing higher standards of achievement for promotion from grade to grade and admission to college by means of more rigorous testing and homework;

- lengthening the school day and school year;

- raising teachers' salaries, especially through merit pay, in order to attract and hold more capable people in the profession; and

- upgrading the qualifications of teachers by testing their competence and improving the quality of their own education.

It is not surprising that the reports of a National Task Force on Education for Economic Growth, or a Business–Higher Education Forum, or a National Science Board Commission, should stress science, mathematics, and computer literacy in their efforts to promote greater economic productivity. And the publicity given to the report of the President's National Commission on Excellence in Education reinforced the economic crisis heralded in its opening sentence: "Our nation is at risk. Our once unchallenged preeminence in commerce, industry, science, and technological innovation is being overtaken by competitors throughout the world." This is attention-getting and newsworthy, especially in the present state of the economy and politics. But it overshadowed what came along in a little-noticed later section of the report:

> Our concern, however, goes well beyond matters such as industry and commerce. It also includes the intellectual, moral, and spiritual strengths of our people which knit together the very fabric or our society.... A high level of shared education is essential to a free, democratic society and to the fostering of a common culture, especially in a country that prides itself on pluralism and individual freedom.

Something similar happened to the Twentieth Century Fund's Task Force. A front-page story in the *New York Times* played up its recommendation that federal funds appropriated for bilingual education should be shifted to English-teaching for non–English-speaking children. Little attention was given to the recommendation of a core program for *all* students that included a "knowledge of civics, or what Aristotle called the education of the citizenry in the spirit of the polity." And less attention was given to the civic reasons that the Task Force stressed the primacy of English:

> Our political democracy rests on our conviction that each citizen should have the capacity to participate fully in our political life; to read newspapers, magazines, and books; to bring a critical intelligence to television and radio; to be capable of resisting emotional manipulation and of setting events within their historical perspective; to express ideas and opinions about public affairs; and to vote thoughtfully—all activities that call for literacy in English.

The Report of the Carnegie Foundation for the Advancement of Teaching went even further to say that the mastery of reading, speaking, and writing the English language is the "first and most essential goal of education." When it came to recommending a mandatory common core curriculum for all high-school students, English was joined with history (of the United States and of world civilization) and civics (government) as the dominant elements of education required for citizenship. By comparison, mathematics and science were regarded as subjects for specialists.

The country has before it numerous structural proposals for

educational reform dealing with academic excellence and quality, asking for higher standards and requirements, and hoping for increasing financing. Perhaps all this is to the good. Yet too few influential voices are asking for a serious reexamination of the historic ideas and values of American citizenship or a reassessment of the function of the schools in shaping the civic values, knowledge, and skills of American citizens in the coming decades.

Just before World War I, William James called for "a moral equivalent" to war: "The only thing needed henceforward is to inflame the civic temper as past history has inflamed the military temper." As we look toward the bicentennial years of the Constitution and Bill of Rights, what is missing in our current debates over educational reform is how best to *inform* as well as inflame the civic temper.

Why is it so difficult to inform, let alone inflame, the civic temper in American schools? We must look beyond the blandness that often characterizes the teaching and the textbooks in social studies. The real reason has something to do with the multiplicity of private agendas constantly pressing on the schools through the dissident pluralisms of special-interest groups. I fully recognize the positive effect that voluntary groups can have on democracy when they act as mediating agents. But some groups may also harm the educational process and the common civic purpose by seeking to impose their particular views on the schools, identifying their own values with the public good.

In the past, such groups have spoken in many tongues. Since 1980, the most outspoken and clamorous voices have carried three general messages to the schools:

● *Technological literacy.* I have already mentioned the powerful demands placed on the schools to develop the technological literacy of their students to meet the economic challenges of an electronic age. "High-tech" is most attractive to business, industry, and labor as well as to the high-school whiz kids who astound their elders with their computer savvy. But questions of ethical or political values of privacy, justice, due process, freedom, and responsibility seem to be unable to compete with enthusiasts who want to make sure that computers are easily available.

● *Consumer sovereignty and family choice.* A second message, proclaimed with growing fervor by conservative political voices, can be summed up in the term "consumer sovereignty." It preaches the evils of government regulation in education, the advantages of private schools over public schools, and the priority of parental rights to seek the educational values they prefer for their children through such means of public funding as vouchers and tuition tax credits. An article in the *National Review* makes clear what conservatives should concentrate on during the next ten years:

There are many things wrong with the schools, from overly powerful teachers' unions to look-say methods of reading instruction. But the thread that links them all is government monopoly. It is this monopoly that forces parents to enroll their children in schools that they know are not as good as they used to be, not nearly as good as local private schools, and often brazenly hostile to the parents' religious and moral convictions.

In this outlook the importance of civic learning or education for citizenship is largely ignored. Such concepts as justice, equality, civic participation, and public responsibility seem to have been subordinated to the values of the marketplace of educational competition.

● *Traditional religious values.* A third set of voices does give the teaching of values top priority for public schools as well as private. They claim to speak for "those traditional moral and religious values that made our nation great." For example, Mel and Norma Gabler in Texas aggressively advocate restoring to the public school curriculum and to textbooks the basic Judeo-Christian "values upon which our nation was founded." For years, they have criticized and attacked those textbooks that they claim undermine such traditional values as "monogamous families, anti-homosexuality, anti-abortion, American patriotism, morality, conservative views, teaching of honesty, obeying laws, changing bad laws through a legal process, etc."

When teachers or administrators in the public schools are faced with arguments like these—or those of the Moral Majority, the Eagle Forum, the Christian Broadcasting Network, or Scientific Creationists—they often become defensive, plead academic freedom, or try to ignore the whole thing. Ignoring the problem has become more difficult, however, since representatives of People for the American Way have challenged the Gablers at the hearings of the Texas Textbook Committee and insisted upon defending textbooks that the Gablers criticize. So these influential hearings have become more of an arena for two-way debate and discussion than simply a unilateral critique by those who object to some of the values expressed in some textbooks.

But how are the educators of Texas and the textbook publishers going to distinguish between the testimony of the Gablers and that of People for the American Way? Both profess belief in and commitment to basic "American" values.

My answer is that we should try to change the terms of the public debate. The diverse claims on the schools should be judged on the basis of their contribution to the prime purpose of universal education: preparation for citizenship in our pluralist democracy. We should try to develop an intellectual framework about civic values by means of which citizens can better judge the policies put forward by business or labor groups, political parties, religious groups, or other special-interest groups. I believe that the primary focus of schools should be on the civic values that pertain to the public life of the political community and not on the private values of religious belief or personal lifestyles that individuals and groups should be free to choose—but not to impose on others, especially not by legislation or government regulation.

I believe that the educational profession should reaffirm the priority of its historic mandate to educate for citizenship and to do so by promoting the most careful scholarly study of the basic historic concepts, principles, and values underlying the constitutional order, about which there has been controversy since the Constitutional Convention of 1787, and by giving fair treatment to conflicting views as revealed in current controversies over the function of government and the meaning of constitutional principles for today.

I have in mind the ideas and values on which our institutions of self-government presumably rest: those ideas that promote cohesion and unity in a democratic political community—justice, equality, legitimate authority, civic participation, and personal obligation for the public good; and those ideas that promote pluralism and individualism in a democratic political community—freedom, diversity, privacy, due process, and human rights. There is tension, sometimes conflict, between these sets of ideas. But I believe that schools should try to promote a realistic understanding of their meaning—in the past and in the present, in theory and in practice.

Such efforts in the schools should include not only the curriculum, textbooks, and formal teaching in the classroom but also the whole range of learning experiences "taught" by the governance and environment of the school itself. This is often called by educators the hidden curriculum. It involves the whole environment of social interaction among teachers, students, administrators, parents, and community agencies.

One of the reasons to speak of "civic *learning*," in preference to the more common term, "civic *education*," is to embrace the multiple meanings of the word *learning*. It is to convey the importance of knowledge and scholarship, as historians do when they refer to the classical revival of learning in and before the Renaissance of European civilization, or as scholars do when they refer to universities as custodians of the world of learning. Learning alludes to a corpus of knowledge and scholarship that informs, stimulates, and challenges the highest reaches of the intellectual, moral, and creative talents of humankind. And civic learning receives sustenance from the primary disciplines of knowledge in philosophy, history, literature, and the humanities as well as in political science, the other social sciences, law, and jurisprudence. But learning also embraces the whole range of experience whereby individuals acquire not only knowledge and intellectual skills but also the values, beliefs, attitudes, commitments, and motivations that affect their learning abilities and achievement at different ages and stages of their development and in the different contexts of family and group life.

The National Assessment of Educational Progress makes it clear that there are regional differences of achievement on tests of political knowledge. Students in the Northeast and Central regions typically perform above the national average level, those in the Southeast below, and those in the West at about the average. The suburban regions around the large urban centers perform above, and the big cities below, the national average. The affluent sections of urban areas perform above, and disadvantaged sections of urban areas perform below, the average. While there is little difference between the sexes, the level of higher education attained by parents proves to be an advantage to their children. Taken at face value, such a finding should mean that the longer the present generation of children and youths stay in school, the better chances *their* children will have of doing well on such tests.

While these test scores show the influence of social class and family background on achievement in school, evidence from political socialization studies shows that schooling in civic education has more influence on the political motivation and knowledge of lower-class children than it does on middle-class children. Civic learning for non-college-bound children is therefore even more important than it is for the affluent children of middle-class and professional families bound for college, who will at least have some chance to acquire more political knowledge and participation skills later. But civic learning is likely to be most important of all for the "unincorporated" youth who have historically been blocked from the mainstream of American political and social life: the disadvantaged minorities, the blacks, and other ethnic and racial groups.

This is all the more reason that children and youths should not be separated from one another into special ability tracks for civic education, or into segregated private, parochial, or public schools. It is extremely important for pluralistic groups to learn about and from each other through study and participation together. Schools should be training grounds for acquiring the sense of community that will hold the political system together. Political efficacy should be shared as widely as possible, if the democratic system is to work. If the goal for every minority youth is a "high school diploma in one hand and a voter registration card in the other," the task of the schools is to be sure that the diploma and the voter registration card rest on the best political knowledge and skill that can be mustered for their learning.

There is no easy road to educational reform. The best first step is for the academic community and the educational profession to reaffirm their commitment to the public purposes of schooling and to the civic goals of universal education. And they must try to convince the public that education is primarily a public good rather than a means to private fulfillment.

In the midst of the cacophony of optimistic hopes and pessimistic fears of what the age of high technology will do to us, it is not only the economic or military prowess of the country that is at risk. The very meaning of our lexicon of democratic values is at risk. To preserve these values will require the combined efforts of state and federal authorities, the civic-minded wings of the legal profession, the scholarly and public-service professions, the political parties, labor and business, the media, the independent sector, the civil rights and civil liberties organizations, women's and ethnic groups, and, not least, students themselves.

They should urge schools to revitalize their civic purpose, including the freedom of teachers to deal in a scholarly way with the controversies that swirl around matters that affect the constitutional order and public policy. The road to civic learning in a pluralist democracy cannot be built primarily on computer sovereignty or consumer sovereignty or family sovereignty. It must be built on citizen sovereignty.

The Dilemma
of the Technocrats

MARK E. KANN

*"If you look into your own mind, which are you, Don
Quixote or Sancho Panza? Almost certainly you are
both. There is one part of you that wishes to be a hero
or a saint, but another part of you is a little fat man
who sees very clearly the advantage of staying alive
with a whole skin."* —George Orwell

Orwell's early writings were centered on what he called the
common decency of the poorer classes. He found camaraderie
among the down and out, courage and cooperation in communities of unemployed miners, friendship and caring in villages,
some moral integrity amidst the lower middle class, and equality of
sacrifice by peasants fighting the Spanish Civil War.

He defined common decency as a combination of individual
autonomy and interpersonal community that could be found in agrarian
villages and working-class neighborhoods. It characterized people
who cultivated their own gardens, enjoyed family life, and strove to
earn enough money to live well without becoming dependent on
material things or on those who monopolize them. Orwell had more
respect for the shopkeeper who was responsible to his neighbors than
for the intellectual who claimed all humankind as his clientele; he
was more enamored of trade unions as brotherhoods than as instruments for winning wage and benefit packages. Common decency, in
this sense, was rooted in a pre–World War I, small-town conservativism.

Orwell also used common decency to connote a mixture of
intelligence and public service. He preferred the passive virtues of
village life, but he was drawn into the larger political arena by
overriding social issues. Especially after Spain, he saw a need to
harness intelligence to truth rather than to political expedience, and
he used the printed and spoken word to communicate that truth. He
was not certain that truth would make people free, but he knew it
would debunk and detoxify the lies that most certainly would make
people slaves. Thus, Orwell often sacrificed his literary aspirations to
compete in the marketplace of ideas, less from personal desire than

from an abiding sense of civic duty. Common decency here was a form of committed liberal citizenship.

Finally, Orwell infused common decency with a radical egalitarianism more akin to Kantian respect for people as ends in themselves than to Marxist visions of material equality. He was most struck in Spain by the "mental atmosphere" in which "many of the normal motives of civilized life—snobbishness, money-grubbing, fear of boss, etc.—had simply ceased to exist." He was intrigued by the democratization of Spanish militias that suspended the symbols of rank to respect the voice of the rank and file. He felt attached to the anarcho-syndicalism that legitimated decentralized control by the workers. From this angle, common decency was a form of radical socialist morality.

These elements of common decency do not make up a coherent whole, and George Woodcock has commented that Orwell was "a man of contradictions." In part, the contradictions were a function of Orwell's lack of rigor and of his changing commitments with the passage of time. In part, however, they were also rooted in a historical era when common decency was everywhere being destroyed.

Conservativism was becoming fascism, subordinating organic relations between individuals and communities to a militarized, jingoistic social order. Liberalism was being overwhelmed by capitalism, replacing thoughtfulness and citizenship with material commodities and self-aggrandizement. Socialism was being transformed into communism, devaluing egalitarianism in the name of political centralization, economic rationality, and Soviet interests. The gravest danger in this upside-down world was that ideology and technology were being used to win popular consent. Ideology had become an attempt to mobilize common decency, through deception, in the service of domination: peace is served by war, freedom requires obedience, and justice is won through injustice. The most pervasive lie was that industrial "progress" is necessarily benign.

Orwell's dilemma was that he could find no social force willing to mobilize against such domination. By the time he wrote *Nineteen Eighty-Four* he preserved at best a diminishing faith in "the proles." He did not believe that members of his own class were a serious force against domination. The middle class nurtured a sense of independence, civic pride, and social justice, but it also transmitted the manners, language, and skills that promised material success only if one served as an engineer, a technician, or a manager for dominant elites.

The contest between common decency and domination was different in the United States, where the ideology of material progress and the social mobility that attended economic growth have muted class distinctions. Certainly, farmers' rebellions against the Eastern

Establishment and early labor struggles for worker control at the end of the nineteenth century represented attempts by the poorer classes to preserve their autonomy, community, citizenship, and equality. But the main drama, if only because it is the abiding drama, has been played out within the class of Americans who labor primarily with their minds.

The marriage between science and technology was consummated at the turn of our century by the formation of a professional and managerial class. This new middle class included, according to Barbara and John Ehrenreich, scientists, engineers, teachers, social workers, writers, accountants, lower-level and middle-level managers and administrators, and so on. Its identifying trait was a claim of expertise in applying scientific knowledge and developing problem-solving techniques. Its social legitimacy was based on its promise to replace politics with intelligence—by eliminating political bosses in American cities, for example, replacing them with meritocratic civil service systems. In practice, however, the professional-managerial class internalized the contradiction between common decency and domination by virtue of its location in the political economy.

This new middle class has codified its own version of common decency in relation to professional ethics. "The professional," writes David Smith, "best serves humanity by preaching what he practices: dispassionate appraisal, skeptical rationality, a respect for facts and figures, and a steady devotion to working a personal synthesis of professional competence and humane values." In a sense, the professional is a small-town conservative who demands the individual autonomy necessary for cultivating his specialty within a community of like-minded professionals. He is also a committed liberal who seeks truth through theory founded on a separation of fact and fiction, tested in the marketplace of ideas. He may even be a radical egalitarian dedicated to serving his clients, countrymen, and humanity rather than special interests or particular classes. Ideally, the professional contributes to "the method of cooperative intelligence" that John Dewey pinpointed as the most advanced means of improving people's lives by facilitating solutions of popular problems. But while professionals may invest considerable sincerity in the common decency written into their codes of ethics, their historical practice has been to lend support to those with power.

This professional complicity has taken three forms. First, professionals have organized so as to guarantee their own autonomy at the public's expense. Their associations demand the right to accredit members and related institutions, to legitimate or delegitimate particular perspectives, and to regulate relations with clients. In the process, they exclude the public from their discussions and ridicule, as Christopher Lasch put it, "popular traditions of self-help as backward and unscientific." The "helping professions," for instance, promise to empower people to control their own lives but often serve as agents of conformity. More power to the professions has meant less power to citizens.

Second, professionals often occupy the middle-management positions in big government, big business, and other big bureaucracies that centralize power in order to regulate human conduct. Such professionals employ their particular expertise to eliminate spontaneity from the behavior of subordinates. They pioneer the time-and-motion studies intended to make labor more efficient and predictable. They administer the labor-saving technologies that make human hands expendable, leaving it to other professional managers to administer jobs programs, retraining, and unemployment. The result has been greater sophistication in the mechanics of social control of subordinate populations. They develop the human relations, public relations, and industrial psychologies that harmonize centralized planning and regularize patterns of consumption. In short, professionals as managers keep the bureaucratic engines humming by ensuring that the human machinery is well oiled.

Third, professionals build a mystique around their monopoly of problem-solving techniques and fancy themselves able to transcend the limits of their particular specialties. Professional economists, for example, assume the mantle of environmental expertise by translating air and water into commodities. Professional businessmen able to solve problems that have to do with generating profits often move into political positions as advisors or cabinet officers. The logic of such generalizable expertise points toward an inegalitarian society in which experts rule and amateurs passively obey.

The contradiction between professional ethics and managerial practice was neatly capsulized in the Pentagon papers episode. The same professionals who produced and publicized this remarkable document had also lent their expertise to managing the lies and distortions that the document exposed. For most of the present century, the new middle class has been able to manufacture myths that reinforce the pattern of class rule, such as the myth that science will solve the problems generated by technology. The rise of the New Left in the 1960s, however, signaled that the offspring of the new middle class were finding the myths less than credible.

The New Left student movement drew its recruits primarily from the sons and daughters of the professional-managerial class, and the students implicitly reflected their parents' fears that professional prerogatives were being eroded by an increasingly managed society. Implicitly, the students defended the norms of common decency inherent in scientific enterprise and professional ethics. They demanded more individual autonomy and community but less subordination to Cold War militarism. They affirmed the search for knowledge and social commitment. They identified with the poor, minorities, workers, and then women, whom they saw as subordinated rather than served by multinational corporations. The immediate symbol of contradiction was, to them, the university. Its announced code of ethics was independent thought, liberal citizenship, and democratic ideals, but its historical practice was to train the next generation of technocrats and managers to work for the Establishment. At least initially, the students stood up for the common decency that their professional parents could no longer maintain.

Consider the plight of physical scientists. At the turn of the century, figures such as Henry Rowland and George Ellery Hale made "a plea for pure science" to defend basic scientific enterprise against those who saw utility only in its technological applications. After World War I, Hale joined with other noted scientists to establish the National Research Council, to secure private funds for basic and applied research, to knit together donors, administrators, and recipients to defend research autonomy, and to dispense funds according to scientific promise rather than special interest. The drive for a national endowment failed. Scientific leaders could not convince donors of the need for basic research. Science soon became Big Science, requiring huge capital outlays available only from the national treasury. With World War II and the Cold War, the federal government's research dollars were based on military priorities.

Rank-and-file researchers thus became more and more a managed population in the postwar years. Individual researchers were assimilated into teams, in national laboratories such as Los Alamos, Livermore, and Brookhaven. The teams became penetrated communities, organized and directed by a scientific establishment that rationalized the laboratory, secured funds for it, linked it to government priorities, advised national policymakers, and even served in federal agencies. Individual scientists found it increasingly difficult to defend their professional autonomy and communities, their critical civic commitments, or their sense of public service against the managerialism of their own class.

Scientists were compensated for their loss of professional prerogatives by newfound wealth, power, and status. Large numbers of young people flocked into expanding science programs, inflating the ranks of Ph.D.s entering the job market. The launching of the new U.S. space program in the wake of Sputnik not only enhanced the material prospects of the profession; it also fueled the myth that Kennedy's New Frontier could be pioneered by "the best and brightest" to produce a new era of progress.

The road to a better future, however, was soon marked with barriers. Government support for research, especially basic research, ebbed during the mid 1960s. New Ph.D.s sometimes faced a choice between working as overeducated technicians or being unemployed. Meanwhile, the scientific profession was infected by the civil rights and antiwar movements, producing radical caucuses that criticized the military priorities of the establishment. The in-house critics of the scientific community argued, for example, that the search for knowledge should be allied with a clean environment and with occupational safety and health rather than with the corporate bottom line. By the end of the 1960s, many physical scientists and their political allies felt that professional common decency must overcome the domination of government by corporations.

A similar story can be told about other segments of the new middle class. The helping professions became more crowded, more bureaucratic, more tied to external funding and regulation, and more distrusted by the public. Professionals educated to serve the welfare state found that their jobs were vulnerable to an impending taxpayer revolt. Professionals employed in the corporate world discovered that oligopolies gave higher priority to guaranteed profits than to intellectual risk, innovation, or efficiency. They learned that big business could show more interest in using professionals to control and break labor unions, or to manage the distribution of pain caused by runaway shops and shutdowns, than in tapping the professionals' more benign areas of expertise. The leverage of professionals was diminishing, moreover, as virtually every occupational category laid claim to professional credentials, thereby diffusing professional authority. After more than a half century of dramatic growth, the new middle class is recognizing its dispensability as a professional class. It is realizing that it is either a creature of upper management or an overt agent of social control.

The New Left was a political force only as long as it spoke to this dispensability; it soon fragmented, however, along lines unacceptable to the new middle class. The counterculture preached a return to nature and spontaneity that required the elimination of science and technology. Few members of the new middle class went along with this plea for class suicide. The Leninist sects that took over New Left organizations were centralized, quasi-militaristic units, but professionals would not support a left-wing version of the same managerialism that had made them dispensable. With little support from its own class, the New Left withered away.

Professionalism has remained on the defensive since the 1970s. Its crisis was exacerbated by a nuclear arms race that required the sacrifice of individual autonomy to national security, by political scandals that announced how little dominant elites cared for truth or social commitment, and by an economic recession that shattered the illusion of middle-class security. But some segments of the professional citizenry have drawn on an ingrained sense of autonomy, efficacy, and service to defend common decency against managerial domination.

Harry C. Boyte estimates that more than 20 million Americans were active in grass-roots citizens' groups during the 1970s. Peter Clecak interprets this activism as the search of citizens for their own "piece of social justice." These activists want to defend individual autonomy by supporting community traditions and offering neighborhood cooperation as an alternative to centralized power. Their efforts to sustain small-town virtues require a working notion of liberal citizenship, and normally passive residents are asked to commit themselves to active political participation. They argue that a revitalized local politics should empower citizens to take greater control over their own destinies and to repoliticize issues too often decided behind the veil of expertise.

Some of this new populist momentum came from the farmers, workers, minorities, and low-income Americans. But the most widespread political mobilizations were rooted in the concerns of the new middle class—environmentalism, feminism, and the peace movement. Unlike reformers from the Progressive Era or the New Deal, the new populists of the 1970s and 1980s do not represent an ascendant class staking out the ground of political leadership for the future. They represent instead an embattled class that is being forced to seek out allies among the poorer classes that have grown skeptical of their claims to expertise. The new populists have learned that the price of interclass alliance is recognition of the autonomy, intelligence and needs of the "amateurs."

Technical solutions to environmental degradation, for instance, will be unacceptable to labor unless they account for the work environment and possible job displacement. Programs of women's studies in universities or advocates of the Equal Rights Amendment will be isolated unless they are connected to the concerns of lower-class minority women who face, say, the high costs of public transportation. Peace proponents will have to speak about conversion rather than outright closure of defense plants, lest they alienate people fearful of filling the welfare rolls at a time of diminishing benefits. In short, the new populists cannot simply reclaim their own professional prerogatives without renegotiating the social contract between specialized expertise and generalized citizenship.

The new populists are now engaged in this process of renegotiation. They draw on the expertise of groups such as Physicians for Social Responsibility, or publications such as the *Bulletin of Atomic Scientists,* to provide alternatives to establishment views on technical issues and to help the public develop scientific literacy. They look back to Saul Alinsky, whose methods of community organizing were codified into "rules for radicals," and they look forward as Derek

Shearer says, to using "the technology of modern electoral politics: computer-aided voter targeting, direct mail literature appeals, and Big League fundraising." The advent of the home computer has afforded them an inexpensive instrument for mobilizing people to compete with established groups and parties. The new populists are also investigating ways to enhance control by citizens and the community by adapting decentralized technologies such as solar energy and cable television. They generally invest their hopes in modern science and technology as sources for revitalizing American political thought and action.

The new populists usually recognize that their own professional pasts have generated a barrier of public cynicism. When science and technology are deemed politically important, it is the professional-managerial class that ends up at the control panels and sends its children off to computer camps. When new avenues of public participation are celebrated, it is usually the efficacious and resourceful educated class that takes advantage of opportunities for civic power. And as long as the mystique of expertise perpetuates itself, the new populists will be tempted to assume the contradictory role as managers of citizen activism. It remains to be seen whether the leaders of the new populists prefigure a more egalitarian future.

Whether the new populists can bear this burden is uncertain, but at least one attempt to fortify public support is emerging. My interviews with the new populists who now run the city government of Santa Monica, California, suggest that the meaning of professionalism may be undergoing a significant revision. Professionals no longer stand above politics but consciously lend their expertise to facilitate democratic discourse and action. Thus, the city manager may use scientific survey techniques to enhance and legitimate the public's voice or to prepare legal briefs that will help citizens understand the possibilities for testing the limits of state and federal laws. The city planning commissioner may translate the jargon of zoning into a vernacular that allows citizens to participate in determining the future of their community. At the moment, these new populist professionals are somewhat torn between career affiliations that demand that they rise above politics to manage society and a politics that requires them to facilitate popular self-management. The tension persists and they have yet to overcome public distrust.

George Orwell had little faith in the politics of what he called "the new indeterminate class of skilled workers, technical experts, scientists, architects, and journalists, the people who feel at home in the radio and ferro-concrete age." He believed that this new class would compromise its common decency to profit from directing big wars, governments, and corporations and that its more socially committed members, such as Winston Smith, would become victims of their own technology.

This class, however, is no longer new or ascending in the United States, and its interests no longer bind it so tightly to the forces of domination. Somewhat on the defensive, it may have to choose between a politicized notion of professionalism that links it to the poorer classes and the idea of service as the creatures and agents of managerialism. Conceivably, the centralization of the "ferro-concrete age" will be followed by the democratic politics of postmodernity. The new populism is an early, faltering, and ambiguous attempt to mediate this future.

Will Orwell be proven wrong? I recently attended a lecture by Jacobo Timerman, who argued that Orwell's pessimism was profoundly flawed; it could not account for the courage of human beings to resist the forces of domination. I imagined Orwell, "a man of contradictions," agreeing with Timerman, because it was Orwell who could write even in the desperate year 1941, "When it comes to the pinch, human beings are heroic." Perhaps when the professional-managerial class feels more pain, it can join with Orwell and choose common decency and democratic citizenship.

American Unions Face High Technology

Jack Golodner

*"Since changes are going on anyway, the great thing is
to learn enough about them so that we be able to lay
hold of them and turn them in the direction of our
desires. Conditions and events are neither to be fled
from nor passively acquiesced in; they are to be
utilized and directed."*   —John Dewey
Reconstruction in Philosophy

"To change and to improve are two different things."
—German proverb

At the time of the Luddites, new labor-saving technologies were
being introduced in the textile industry in the midst of poor
overall national economic performance. Textile workers, already facing massive unemployment, were threatened with the permanent loss of their jobs and with lives of abject poverty. Confronted
by an indifferent government and myopic manufacturers and lacking
any alternative means of winning back their jobs or sharing in the
new wealth promised by increased productivity, they lashed out in
their frustration. In part, they succeeded, for as Rosalind Williams,
writing in *Technology Illustrated,* noted,

> Luddism . . . did buy them time. . . . Because of the threat of machine
> breaking, employers thought twice before investing in new devices
> or cutting wages. Nor were the workers' tactics politically futile.
> Luddism was a significant factor behind the British Government's
> decision to repeal the continental blockade in June 1812, which
> resulted in rapid economic improvement.

The point is not that the destruction of machinery is proper.
The lesson of the Luddite experience is that the widespread introduction of labor-saving technologies cannot be carried out successfully
without concern for the social consequences—in other words, it can
be accepted, but only grudgingly and suspiciously.

Today, the problems that attend the spread of high technology
are as difficult to address, if not more difficult, than in early nineteenth-
century England. The introduction of computers and microchips into

almost every facet of human activity requires nothing less than a top-to-bottom redefinition of work and its place in human endeavor. Unfortunately, as in nineteenth-century England, our government is largely indifferent. The entire power structure seems complacent and uncomprehending.

In the past, when work-abolishing mechanisms were introduced, they were usually confined to particular sectors of the economy or to specific functions within a sector or industry. Workers displaced from one area, therefore, could with some adjustments move to another, less affected technology. Just over a hundred years ago, more than half the U.S. work force were engaged in agriculture. Today, less than 3 percent are so occupied, though American farms are capable of feeding a good portion of the world. Where did the people go? They went into manufacturing, where they eventually helped make American factories the most productive in the world. And when technological innovations displaced them from the factories, the people moved on to the service industries. Today, two thirds of the American work force are engaged in white-collar and service occupations. But now these jobs are being computerized, robotized, and oftentimes eliminated by sophisticated marvels of artificial intelligence. Today, *every* sector of the economy is affected by labor-abolishing innovations. The cumulative effect could be devastating.

Where will the next generation of new jobs come from?

What kind of jobs will they be?

How rewarding will they be?

Under what conditions will they be performed?

What are the consequences for human growth and freedom?

In the predictable future, the new jobs will surely not come from the high-tech industries. David Kearns, chief executive officer of Xerox, reminds us that "high tech, by nature, is not labor intensive and will probably never generate anywhere near the number of jobs in basic industries and services." In 1979, high-tech industries—drugs, computers, aircraft, electronic components, chemicals, communications, and so on—employed 3 million workers, or 3 percent of the nonagricultural work force. That percentage is expected to rise only to 4 percent by 1993. Obviously, the new glamour industries will not be able to provide work for those displaced from the old smokestack sector, much less help us employ the 6 million people who are now unemployed for other than frictional reasons. Indeed, more jobs will be lost to technological displacement than will be gained in the advanced technology fields during the coming years. The communities that are losing industrial jobs are not likely to be the same ones that attract high technology. While most of the new jobs will not be as demanding of skill or as creative or pay nearly as well as the old ones, a few new jobs will be highly productive, highly responsible, highly skilled, and highly paid. Many experts predict that the latter development could eliminate the middle-class worker and change the character of the American work force to one of extreme haves and have-nots.

For some time now, the new technology and higher productivity have not stopped a trend toward greater and greater unemployment but, indeed, have exacerbated it. Since 1950, each major recession has left the United States with a higher rate of unemployment, and since 1974, the number of Americans living in poverty has increased from 23 million to 30 million.

Wassily Leontief, Harvard's Nobel laureate in economics, has looked at the past, when new technologies spurred the expansion of employment and better standards of living. But he says that history won't necessarily repeat itself. "Labor will become less and less important," he says. "More and more workers will be replaced by machines." He claims that conditions facing today's unemployed are less like those of farm workers who moved to manufacturing and more like the plight of farm horses returned to pasture.

Obsolete—like the farm horse. That could be the fate of millions who will have to deal with growing periods of little or no income, and worse, no place to satisfy their need for effort and creation. This is hardly fertile territory for human growth and freedom.

What is labor's answer?

If we are to produce more and more wealth, then labor insists that commitments be made *now* to ensure that this new wealth will be shared. We must resist a know-nothing drift toward a society in which the few work at significant, challenging jobs and receive large incomes, while the many live on the margin of despair—most of them unemployed, their skills and talents unused, plodding through boring jobs or endless days of no work at all, struggling to subsist.

The wealth that will be derived from greater productivity can be redistributed in the form of income, in time, or—better still— through a combination of both. By maintaining and even increasing the incomes of American workers, we ensure that there will be sufficient demand to sustain the higher levels of output made possible by greater productivity without causing even more reductions in the work force. Higher incomes for individuals also mean more government revenues to support improvements in those activities that only the government can perform—maintaining and expanding our transportation and social-services systems. And expanding these activities as wealth increased would also provide worthwhile employment.

By reducing the working hours of those employed, we make it possible to share the jobs, if jobs do indeed prove to be scarce, as now seems to be the case. Leontief points to the fact that the working hours of most Americans steadily declined until World War II, when, with slight variation, they froze at forty hours a week. Though a growing number of people postponed entry into the work force by extending their stay at school, this was not nearly sufficient to counter the ratcheting upward of unemployment from the late 1940s. We must renew the historic effort to increase the nonjob time of the American worker.

I use the term "nonjob time" advisedly. I do not mean nonwork time. If proper opportunities are provided by our society there will be no absence of work. Given the chance to spend more time away from the job, people will be free to pursue what they know to be their *own* work, whether it be gardening, the arts, continuing education, or development of their physical skills. Erich Fromm warned that a twenty-hour work week would amount to a WPA program for psychiatrists. This would not happen if more were done to use the new wealth made possible by the new technology to provide more opportunities in the arts, education, and recreation.

In recognition of this fact American labor has, since its beginnings, argued for greater educational, cultural, and recreational opportunities. The provision of such opportunities will not only enrich the nonjob time of those who participate but will also create many new jobs in those areas.

The specter of millions of people having nothing to do but saunter and sit and be silly and do nothing is frightening. Equally threatening to the human condition are predictions that, because of high tech, more and more jobs will be monotonous, repetitive, boring.

According to a study by Levin and Rumberger of Stanford University, the total number of new jobs generated in high-tech corporations will be vastly outweighed by the number of jobs generated in other areas. The five occupations expected to produce the most new jobs in the 1980s, for example, are all in low-skilled categories—janitors, nurses' aides, sales clerks, cashiers, waiters, and waitresses. No high-tech occupation even makes the top twenty in total numbers of jobs added to the U.S. economy. While employment for engineers, computer specialists, and other high-tech professionals will grow almost three times as fast as total employment, these occupations will generate only about 7 percent of all new jobs during the rest of this decade.

One must wonder what will happen to the one out of every three entry-level workers holding college degrees in 1990, when the largest share of the job openings may be at fast-food counters.

"The logical end of mechanical progress," George Orwell warned, "is to reduce the human being to something resembling a brain in a bottle." This is the goal toward which we are moving, though we have, of course, no intention of getting there—just as a man who drinks a bottle of whiskey a day does not intend to develop cirrhosis of the liver. "The implied objective of progress," he wrote, "is some frightful sub-human depth of softness and helplessness."

We are all alarmed that test scores in schools are falling. We are told that students today are not motivated to learn new skills and acquire knowledge. Perhaps in some mysterious way yet unknown our young have received a message from the future: Why persevere in studying material that will not be needed? Why develop skills that will not be used?

If job-related work is not to sustain the need of mankind for effort and creation, then society must prepare *now* to offer constructive nonjob activities. If the robots are truly modern man's slaves, they must be slaves not only to an elite, but to all people. Everyone must be freed by these slaves to live as did the citizens of ancient Greece— free to devote their full time to developing their minds and their bodies and to participate in all matters of consequence to the community.

Workers of an earlier age turned to unions in search of freedom from oppressive management, from exploitation, from unsafe and unhealthy working conditions. Today's worker looks upon unions as instruments to achieve the freedom to participate in the decisions that affect their jobs, their communities, their society; to explore their minds and their bodies; to grow and enrich their lives; to move toward a better life, not merely away from an oppressive one. Technology promises this. And workers—through unions—insist that society fulfill this promise.

Lou Harris has reported his finding that most Americans fear that computers will invade their privacy. "Make no mistake about it," he said. "Lying in wait out there as people contemplate the future in the information age are a whole series of wrenches, apprehensions, dislocations and downright potential horrors that they feel are part and parcel of the baggage of the new times." In this the American people are smarter than their leaders in government, who are doing little to protect them against the Big Brother paraphernalia of the high-tech age.

For evidence of what is coming, talk to the workers who have already experienced the introduction of new technology the aim of which is not greater productivity, but an increase in managerial control. Software has been developed for word processors and computers, for example, that can count the number of keys hit in a given period.

There is no reason for the operator of this equipment to want such information, but some American managers routinely use these data to measure productivity. One result is an increased emphasis on hitting keys—whether they are the right keys or not. Errors increase, necessitating correction at a later stage. Stress levels for the operators have been known to increase significantly. Is this progress? Is this innovation that frees the worker to do his or her best?

The computer also makes possible the growing practice of block modeling—a new information technology for classifying individuals on the basis of large quantities of data or large numbers of transactions and relationships. Using this method, for example, corporations can screen out potential employees who—according to their block category—may be inclined to join a union. Block modeling is being used to separate fast-tract from slow-track employees.

What defense does the average employee have against the possible misuse of block modeling? None. Our lawmakers, like the British Parliament before it was awakened by the Luddites, are asleep. So unions have become the protectors of individual rights at the work place by negotiating contracts with clauses that place restrictions on senseless monitoring, require employers at least to notify employees that they are the subjects of block-model analysis and to reveal the methodology as well as the basic data used.

Do I sound like Cassandra? No more than Orwell.

I believe we have the means within our society—through government and through our private associations such as free trade unions—to prevent our worst nightmares from becoming reality. I am hopeful that men and women of learning and influence will wake up and act.

Like Orwell, the labor movement accepts the machine, but suspiciously.

Voluntary Associations and Democratic Governance

DAVID L. SILLS

"There are no countries in which associations are more needed, to prevent the despotism of faction or the arbitrary power of a prince, than those which are democratically constituted." —Alexis de Tocqueville
Democracy in America

It is obvious to the most casual observer that organizations are essential to modern society—in fact, to any society larger or more complex than a face-to-face band of hunters or gatherers. For this reason, organizations have frequently been examined by social scientists, and as frequently criticized. Max Weber saw them in part as manifestations of the inexorable process of rationality, the "iron cage" in which modern man is imprisoned "perhaps until the last ton of fossilized coal is burnt." He also saw them, of course, as social inventions essential to the functioning of a specialized society. Robert Michels, in his 1911 study of political parties, discovered what he called "the iron law of oligarchy," according to which "it is organization which gives birth to domination of the elected over the electors, of the mandatories over the mandators, of the delegates over the delegators." "Who says organization," Michels intoned, "says oligarchy." More recently, the sociologist Philip Selznick has written of "the tragedy of organization": the tendency of organizations to defeat the very purposes for which they were established. Mancur Olson, in *The Logic of Collective Action* (1965), maintains that it is logically impossible for voluntary associations to work the way they are supposed to work, since rational individuals will not voluntarily seek a collective good for a large group. In his *Rise and Decline of Nations* (1982) he maintains that Germany, Japan, and a few other Asian countries have had greater postwar economic growth than the United States and the United Kingdom precisely because they have been less hindered by an infrastructure of voluntary associations. And ever since the publication of *The Organization Man* (1965), by William H. Whyte, Jr., "pop" sociologists have been compelled to point out that large organizations are destroying individuality in America as well as generally messing things up. In spite of these discouraging analyses, however, private organizations—specifically, voluntary associations—are crucial instruments both for combatting tyranny and for keeping us free.

In order to defend this thesis, it is necessary to define voluntary associations, to indicate their extent and membership, to explain the principal organizational processes that influence them, and to spell out the functions that they serve, both for the individuals who belong to them and for society at large. My emphasis throughout is largely—but not entirely—on the United States.

What is a voluntary association? Definitions differ widely, but a voluntary association is generally an organized group of persons formed in order to further some common interest of its members, in which membership is voluntary, neither mandatory nor acquired through birth, and exists independent of the state. Even this broad definition marginally admits some exceptions. Membership in such voluntary associations as labor unions and professional societies may be a condition of employment or professional practice and thus may not be truly voluntary. Membership in a church may be "inherited" from one's parents and, in that sense, not voluntary. Many voluntary associations are subject to state control to the extent that they must be registered, and agencies of the state often create or sponsor voluntary associations in order to achieve their own ends. In spite of these exceptions, however, these three characteristics provide a framework for considering the distinctly sociological importance of voluntary associations.

Although all nonstate, common-purpose organizations with voluntary memberships depend on freedom of association, I will concentrate on organizations that meet two additional criteria. First, the principal activity of most of these organizations is not related to the business of making a living. Second, the volunteer—that is, nonsalaried—members constitute a majority of the participants (as they do not in corporations, universities, or foundations, in which the directors or trustees are in a minority and the employees, faculty, or students constitute the majority). "Spare-time, participatory associations" might be the best description of voluntary associations.

Extent and membership. The data on the number of spare-time, participatory voluntary associations in modern societies, and on the proportion of the population that belongs to them, are scattered, and comparisons among studies are made difficult by differences in definitions and in research objectives. The 1984 edition of the Gale *Encyclopedia of Associations* describes more than 17,500 associations in the United States, but it obviously misses most local organizations. The sociologist Albert J. Reiss, Jr., has suggested that there may be more voluntary associations in the United States than there are people.

Since Tocqueville made his famous observation in *Democracy in America* (1835)—"In no country of the world has the principle of association been more successfully used or applied to a greater multitude of objects than in America"—and James Bryce made his in *The American Commonwealth* (1888)—"[They are] created, extended, and worked in the United States more quickly and effectively than in any other country"—the United States has often been described by observers as "a nation of joiners." In 1911 Max Weber expressed the opinion, "What is, in qualitative terms, the association-land par excellence? Without doubt America."

In spite of the number of such remarks, and without denying the central importance of voluntary associations in American life, perhaps too much emphasis has been placed upon the scope and uniqueness of the American pattern. Max Weber, in 1911, found some 300 associations in a German city of 30,000 merely by consulting the city directory. The observation is frequently made that voluntary associations are most common in America and the Protestant countries of Europe, and indeed the contrast between the Protestant countries and the Roman Catholic countries is quite marked. Yet the prevalence of voluntary associations in such different countries as Japan, Nigeria, Ghana, Thailand, and Israel indicates that the pattern is by no means confined to Protestant countries.

But only in America has sufficient research been undertaken to provide a reasonably accurate picture of the extent of participation in voluntary associations. Depending upon the location of the study and the definition used, it can be said that between 35 and 55 percent of American adults belong to one or more voluntary associations. Membership is not random: some segments of the population are much more likely to participate than are others. Whites are more likely to participate than are Blacks, Jews more likely than Protestants, and Protestants more likely than Roman Catholics. Urban and rural nonfarm residents are more likely to participate than farm residents, parents more likely than nonparents, and frequent voters more likely than nonvoters. The largest and most consistent differences in participation are those defined by socioeconomic status, whether measured by income, occupation, home ownership, or educational level. A majority of Americans of higher status belong to voluntary associations, and a majority of people of lower status do not.

Organizational processes. There are four processes that are of fundamental importance for an understanding of the way voluntary associations function: institutionalization, minority rule, goal displacement, and goal succession.

In its broadest meaning, institutionalization is the process through which patterns of behavior and expectations of behavior become established. Marriage, the marketplace, and the burying of the dead are all examples of institutions. Institutions are largely unplanned: only after it has been well established is a pattern of behavior called an institution.

Applied to organizations, institutionalization means the often unplanned process that turns a loosely organized group of adherents of an idea or a goal into a formal organization. Since it is a social process, institutionalization can be studied at any point on the time continuum that runs from the emergence of an idea to the death of an organization, perhaps through strangulation by the excessive use of its own rules of procedure. Institutionalization is also a characteristic that distinguishes among voluntary associations at different times. Some voluntary associations have goals and programs that are aimed at the gradual improvement of the existing order. Their members, therefore, bring a relatively low degree of enthusiasm to their participation, and the organizational structure is relatively formal and matter-of-fact. Such highly institutionalized organizations can be called formal organizationlike associations; the American Red Cross is an example. Other voluntary associations have goals and programs that are much more radical and ideological and are more at variance with what the participants believe to be the norms of society. Their members bring a relatively high degree of enthusiasm to their participation, and the organizational structure is likely to be informal and fluid. These less institutionalized organizations can be called social movement-like associations; the Congress of Racial Equality (CORE) is an example.

Voluntary associations exist only in societies in which freedom of association exists. Since such societies are more or less democratic in their ethos and political structures it is generally expected that members will take an active part in the affairs of the association and that democratic procedures will govern its conduct. This expectation is often not met. Although most voluntary associations have constitutions, bylaws, or oral traditions that call for full participation by the members, the "iron law of oligarchy" formulated by the German social scientist Robert Michels generally has greater weight. Most voluntary associations—even those with democratic goals—are run by minorities.

A few years ago I analyzed the explanations for the inactivity of the members and minority rule that researchers had revealed in voluntary associations. Some determinants, I found, are located in the very structure of voluntary associations: their large size, their representative form of internal government, their functional specialization, the heterogenity of their membership, and their degree of institutionalization and formalization. Other determinants are located in the requirements for leadership. Leaders must have certain skills, they

David L. Sills

must have time to devote to the organization, and they must have the necessary temperament.

The sociologist John Lofland in 1981 gave a breathless description of the temperament necessary for the leaders of voluntary associations:

> The concrete work of [running voluntary associations] seems best done by people who are quite social and sociable (who "mix well," as it is phrased), who are content to suffer a mélange of fools patiently and unendingly, who are happy to attend meetings and conferences into infinity, who can quickly master complex and tedious documents produced by manifold agencies (each with "hidden agendas"), who can handle the stress of frequent, nasty surprises sprung by competitors, enemies, or even friendly interest groups, who do not become too rapidly bleary-eyed under the unbroken onrush of words and paper, who thrive on lunch after lunch and dinner after dinner of rubber chicken meals eaten in hotel banquet rooms under harsh fluorescent lights, who do not mind waiting untold hours in officials' reception rooms and government hearing rooms, who find it exhilarating rather than devastating to be challenged hostilely in public by officials and others who doubt one's pleadings, who can remain optimistic in spite of wavering and uncertain support given by members of one's own interest group, who can remain cool and good-humored in the face of frustrating and drawn out and frequently unsuccessful processes of decision-making, who are prepared overall, to give themselves over to the extremely demanding flow of events in the interest group struggle.

Finally, determinants of participation by members and minority rule are found in the nature of organizational activities: the absence of concrete tasks conflicts with the programs of other organizations to which members belong, and the often wide disparity between the reasons members join an organization and the actual activities they are asked to undertake.

This array of determinants is similar to the array that might be developed to demonstrate the impossibility of the full participation of all citizens in any democracy. Since it is impossible for all citizens to play active roles, governments are established, and through the process of election, representatives are selected who speak for their constituencies. In many democracies, however, full participation in even the basic activity of voting for representatives is not even approximated. Voluntary associations should be different from nations, it is often claimed, since their members join of their own free will, and in many instances they do participate at higher rates—say in voting—than do members of national states. But similar social-structural determinants serve to depress full participation.

Since voluntary associations, like all organizations, are established for the purpose of realizing some short-term or long-term goal, the relations of these goals to other aspects of an organization's existence have occupied the attention of many researchers. How goals are established, how decisions are made to determine which activities will serve to realize these goals, how goals are interpreted to the community—these and many other questions are implicit or explicit in research on voluntary associations.

The generic problem of preservation of goals may be stated as follows: In order to accomplish its goals, an organization establishes a set of procedures or means. In the course of following these procedures, however, the subordinates or members to whom authority and functions have been delegated often come to regard the procedures as ends in themselves, rather than as means toward the achievement of organizational goals. As a result, the actual activities of the organization become centered on the proper functioning of organizational procedures, rather than on the achievement of the initial goals. For examples, look around you.

One source of displacement of goals is the desire of active participants to retain power in the organization; in order to do this, they focus their participation on self-serving activities rather than those that will serve the goal of the organization. Perhaps the most striking example is that of leaders of labor unions, who are understandably reluctant to forfeit their status and income in order to go back to the bench.

A second source of goal displacement lies in the strict enforcement of organizational rules and the lavish carrying out of organizational procedures. The sentiments that are developed to buttress the rules and perform the procedures often become more intense than is technically necessary. Following the rules and carrying out the procedures become rituals, ends in themselves.

The informal structures that develop within organizations are a third source of goal displacement. Informally created factions within organizations are not only crucial to the achievement of goals but are also—through the creation of dissent, for example—responsible for a certain amount of goal displacement.

The literature on the two primary pathologies of voluntary associations, the tendency toward oligarchy and the tendency toward displacement of goals, leads to fundamentally pessimistic conclusions. In spite of the best of intentions, the inherent nature of the process of organization is such that minorities will gain control in order to serve their own ends. As a result, the purposes for which the organization is established will be blunted. The two landmarks in this literature illustrate the point. In his *Political Parties* (1911) Michels concluded that the prospects for democracy in social democratic parties and labor unions were dim. Philip Selznick, in *TVA and the*

Grass Roots (1949), concluded that delegation by the TVA of certain phases of its agricultural program to local organizations simply led to the co-opting of these organizations into the policymaking apparatus of the TVA itself—which in turn caused the TVA to be deflected from some of *its* primary goals.

Nevertheless, a recurrent theme in the vast commentary on Michels's iron law of oligarchy is that the law holds true only under certain circumstances, and even when it does hold true, other democratic values may yet be furthered. In a frontal attack on those who see only tragedy in the processes of organization, Alvin Gouldner asserted that "there is every reason to assume that 'the underlying tendencies which are likely to inhibit the democratic process' are just as likely to impair authoritarian rule. . . . There cannot be an iron law of oligarchy . . . unless there is an iron law of democracy." Similarly, the literature of criticism of the concept of goal displacement admits the existence of the pathology but points out that the new goals of an organization may be an improvement upon the old ones.

Functions for individuals. The notion that individuals seek out and join voluntary associations in order to find outlets for their interests is oversimplified, since there is considerable evidence that most individuals join associations only after being urged or invited to do so. Nevertheless, after they become members, it may be presumed that individuals benefit to some extent from the organizations' programs, whether the benefits are the satisfactions of sociability, recreation, service, or political action. Manifest functions of this kind, important as they are, require no explanation here. I will focus instead on two latent functions of participation: social integration and training in organizational skills.

That people interact with others when they participate in voluntary associations is obvious, and the benefits of interaction—easing loneliness, learning norms, acquiring information—are among the most frequent functions of membership for the individuals involved. What is more problematic is whether secondary groups such as voluntary associations serve the same integrative functions as primary groups. Louis Wirth and many others have asserted that the weakness of family and neighborhood ties in modern, or urban, society is compensated for by participation in voluntary associations, but such statements do not constitute proof.

The research evidence from studies of American society to back this claim is inconclusive. There is no doubt that some people achieve familylike satisfactions from participation; this is most true of lodges, fraternal orders, and such self-help associations as Alcoholics Anonymous. In most other associations, however, the segmental, part-time, and functional nature of the activity precludes the development of true primary-group ties. What is much more likely is that people who have satisfactory primary-group ties are more likely to join voluntary associations—a reversal of the direction of causality implied by the integration hypothesis.

One explanation for this is that on the whole we are already an integrated society, in which the family and the neighborhood continue to fulfill their historic functions. Morris Janowitz, for example, challenged Wirth's assertions about the anonymity of urban life by describing how the residents of Chicago—the same city that Wirth studied—are bound to family and neighborhood groups.

Functions for society. There is necessarily some overlap between functions that voluntary associations perform for individuals and those that they perform for society. The training of individuals in organizational skills, for example, not only offers them satisfactions and enables them to advance their careers; it also provides the total society with fresh cadres of leaders who have new perspectives on problems, thereby stimulating social change. The distinction is an important one, however, since a function that may benefit individuals—through opportunities for self-expression, for example—may be detrimental to society—if self-expression interferes with the privacy of others, for example.

I would like to comment on six functions that voluntary associations are said to fulfill for society. As is true of functions for individuals, the evidence that these functions are actually performed is uneven, and the problems of verification are largely unsolved.

The term "secondary groups," as it is applied to voluntary associations, indicates that associations mediate between primary groups and the state. Professional associations, for instance, mediate between their members and the government, especially in such matters as licensing, research funds, and legislation. In these matters mediation shades off into lobbying. On the other hand, professional associations, through their programs of public relations and public information, mediate between their members and the general public. In national churches, both local and special-interest associations mediate between the individual members and the church hierarchy.

Foreign policy, to take another example, is the responsibility of the federal government. Yet private groups such as the Foreign Policy Association and the Council on Foreign Relations are a part of the process. The International Research and Exchanges Board (IREX), a private group that sponsors scholarly exchange programs with the Soviet Union, is at present establishing a voluntary national coordinating agency for nongovernmental groups that have policy-related contacts with the Soviet Union. Few would assert that this is meddling in the affairs of government; it is rather an example of useful mediation.

In pluralistic societies, voluntary associations may serve to integrate minority groups into the national society. Many ethnic associations, such as the National Association for the Advancement of Colored People (NAACP), have been formed for this express purpose. The success of such organizations can, of course, lead to their own disintegration. As the psychologist Kurt Lewin pointed out, with specific reference to American Jewish organizations, "The task of organizing a group which is economically or otherwise underprivileged is seriously hampered by those members whose real goal is to leave the group rather than promote it."

Voluntary associations may serve as a legitimate locus for the affirmation and expression of values, as do patriotic societies and political parties. Consider the situation in East Germany, where it is government policy to quash antiwar and antinuclear protest groups. According to a November 28, 1983, *New York Times* report, the Protestant churches have hesitantly offered the use of their meeting rooms to such groups. Since the churches represent "the only democratic institutions in East Germany," they are serving their historic function of expressing values held strongly by people.

It was noted earlier that the uniqueness of the American pattern of voluntary associations has been overstated by many observers: many other countries have large numbers of active associations. It is difficult, however, to overstate the part played by voluntary associations in the actual business of governing the United States, in the sense of making decisions on policy and of providing services to citizens. Tocqueville was impressed by this: "Wherever at the head of some new undertaking you see the government in France, or a man of rank in England, in the United States you will be sure to find an association." Peter Rossi concluded, on the basis of a study of the leaders of a Midwestern industrial city of about 45,000 that he named Mediana, that this pattern is, if anything, more prevalent today than it was at the time of Tocqueville's visit:

> The most striking characteristic of contemporary cities, compared with the American community in the nineteenth century, is the relative drop in the importance of local government, not only in its relation to state and federal governments but also in its relation to local voluntary associations. To understand what is happening within a contemporary community an investigator cannot confine himself to the official table of organization for municipal government but must add to it a host of voluntary associations which act on behalf of the community and which together with the formal structure of local government form the basic organizational framework of the local community.

In large cities, voluntary associations seem to serve largely as important pressure groups; in medium-sized cities, they virtually run the municipal government. In small towns, the decisionmaking function is filled by families and cliques, leaving to voluntary associations such tasks as raising funds for the library, decorating the plaza, and maintaining the cemeteries. Throughout most of the United States— almost everywhere, in fact, except in large cities—voluntary associations perform the fundamental governmental function of coping with emergencies: sickness is treated in voluntary health centers and hospitals; fires are fought by volunteer fire departments; disaster relief is furnished through the American Red Cross.

The organizational machinery for the licensing of such professionals as lawyers and physicians is generally run by voluntary professional societies. Throughout the federal government, agencies rely upon voluntary associations, not only for research and training services and for carrying out public information campaigns, but also, in many instances, for the actual administration of the agency's program.

Why are voluntary associations so effective at governing? One explanation is provided by a social phenomenon that the sociologist Mark Granovetter calls "the strength of weak ties." He asserts that our acquaintances—"weak ties"—are less likely to be socially involved with one another than are our close friends—our "strong ties." So when we have some form of social contact with an acquaintance, we tap into his or her network of friends—all new to us. This is said to be the strength of weak ties: they enable ideas and influence to be disseminated through a community much more rapidly than do strong ties.

Membership in voluntary associations is segmental and part-time; members are generally acquaintances, not friends. This is the reason voluntary associations are called secondary associations, in contrast to primary groups, such as the family and the social circle. Accordingly, the ideas of a voluntary association, its goals, and its activities become quickly known throughout a community; since most members belong to different social networks, the results of a meeting attended by, say, a dozen people are disseminated among a dozen independent social networks. Although there is little research as yet to support this formulation, there is reason to believe that voluntary associations often achieve their purposes, often get things done, precisely because their members are "only" acquaintances.

Since most voluntary associations are formed for the purpose of bringing about some change or improvement in society, and since most of their efforts have had some measure of success, it follows that the initiation of social change is one of their primary functions

David L. Sills

for society. The limiting case of social change—revolution—demonstrates the point. Historically, revolutions—in France, the United States, Russia, or China—either have been started by voluntary associations or have been directed by them once mass unrest has led to outbursts of violence. The Solidarity movement in Poland is a classic example of a voluntary association that has initiated social change—or it will be if it survives and wins.

In many countries, most of the services that are now assumed to be the responsibility of government were initiated by voluntary associations. This is particularly true of welfare services to the poor, the ill, the orphaned, and the aged, but it is also true of education. In many of the developing countries the present public school system is modeled after schools established by missionary societies, and these societies often continue to sponsor the secondary schools in which many members of the future elite are trained. This pattern is not confined to non-Western societies. The public school system of New York City is the successor to a voluntary association—the Public School Society.

One way that voluntary associations bring about change is by making nuisances of themselves, whether by shouting at public hearings, presenting petitions, or blocking the forward movement of bulldozers—in short, by nagging. This term was recently added to the vocabulary of the social sciences by the sociologist James Beniger, who defines nagging as behavior that takes place when we feel positively toward the object of our nagging behavior, want to change the behavior of these others, find that the recipient is either not attentive to us or not receptive to our suggestions, and lack sufficient power to influence the recipient decisively. This, I submit, defines the behavior of many voluntary associations.

The doctrine of political pluralism asserts that the power of the sovereign state must be balanced by the power of dispersed associations. It is generally said to have originated around the turn of the century in the writings of Otto von Gierke, Émile Durkheim, John Neville Figgis, F. W. Maitland, and others. Years earlier, however, Tocqueville gave classic expression to the notion that the power of the sovereign state is best limited by voluntary associations. The pluralist theory seems to have been validated so far. In spite of all our shortcomings, we have escaped totalitarianism, we have centers of power throughout the country, and we manage to train new leaders to replace the old ones. Most scholars agree that two current social trends will continue. Social and economic power will be extended to new classes of people, and the power of government and large organizations will increase. Only the Bourbons among us will challenge the first trend; only the utopians among us will question the second.

The trick, of course, is to avoid an Orwellian future by continually creating, then sustaining new sources of power in society. As Tocqueville put the problem:

> Among the laws that rule human societies there is one which seems to be more precise and clear than all others. If men are to remain civilized or to become so, the art of associating together must grow and improve in the same ratio in which the equality of conditions is increased.

By stating the relation between power and equality in the form of a scientific law, Tocqueville earned the right to be called one of the first truly modern social scientists. And by calling for an improvement in "the art of associating together," he set the basic agenda for an understanding of the relation of voluntary associations to democratic governance.

The Chosen Many

Lewis H. Lapham

A mong the classified advertisements in one of its December issues the *New York Review of Books* carried the following personal notice:

> *AM I BEAUTIFUL? Some people say so. WASP brunette, 5'8", Ivy-educated, professional success, graceful, competent, serene, nonsmoker, nonbitch, nicely proportioned in body and mind. Too intelligent and alive to be happy with loser or numbskull. If you're civilized, solvent, emotionally available, spiritual, epicurean man over 40, who loves his work and himself, seeking multidimensional companionship with grown-up divorced woman, please write. Shy, conventional, reserved exterior okay, but please no sugar daddies, hunters, hurters, or walking wounded.*

Reading the personal notices in the *New York Review of Books* is like looking at the pictures in *Harper's Bazaar* or *The New Yorker*. They represent the upscale end of the market in sensibility, giving shape and form to the styles of feeling in vogue among people who assume they belong in the best intellectual society. It is an intimidating crowd, well-groomed in its literary dress and deportment. The ads set uncompromising standards: "trim," "intelligent," "caring," "fond of fall walks and classical music," "sense of humor," "rich."

With the author of the notice in hand, we are dealing with a woman of intellectual means. This is a woman who has read William Styron and has been to Paris, who knows the difference between a haiku and a sonnet, who listens to Beethoven string quartets while jogging through Brooklyn, who can pronounce the names of at least three South American poets.

A woman so graceful and serene, accustomed to private showings and opening nights, obviously does not need to concern herself with the seedier sort of professor who reads secondhand books and still smokes, in homage to the defunct Camus, Gauloise cigarettes. Nor does the lady want to have anything to do with weakness, emotional dismemberment, or fumbling in the back seats of taxis. She has had enough of lofts and existentialism, of men who wear their dreary inadequacies like suits of ill-fitting clothes.

Why should such a lady put up with inferior merchandise? Given

This essay first appeared in the *Yale Literary Magazine*, vol. 150, no. 4.

her education, her intelligence, her success, how could she be pleased with a mirror that did not reflect her competent perfection?

Although phrased in slightly more ponderous terms, these same questions appear continually in the earnest and seemingly endless American discussions of leadership. The discussions go forward in the intellectual journals, in the newspapers, at conferences and seminars by anxious sociologists and worried tycoons. It so happened that during the same week that *The New York Review* published the desiderata of the lady at loss for love, *Time* published a companion piece entitled "Job Specs for the Oval Office." Concerned about the prerequisites of leadership, Hedley Donovan, former editor of the magazine, wrote the precise analogue of a personal advertisement for the ideal American president. He listed thirty-one attributes desirable in the country's Prince Charming, in a prose style equally remarkable for its dreaming narcissism.

> We prefer presidents to look like presidents.... The president ought to be an athlete, or at least an outdoorsman ... the president needs presence, dignity, a certain touch of distance, and even mystery ... he must be steady and stable, housing his exceptional combination of gifts within a personality approximately "normal" ... the president needs superior intelligence ... the president needs to be an optimist.... We want the president to be flexible, pragmatic, capable of compromise—also firm, decisive, principled.

It is a wonder that Donovan neglected to mention "civilized," "emotionally available," "epicurean." Perhaps he did not have enough space, or perhaps he thought that "epicurean" might be misunderstood in Omaha. Certainly he wants a man of the world who is capable of "multidimensional companionship" with the electorate. God knows, the electorate has had enough unpleasant experiences with presidents who turned out to be "sugar daddies, hunters, hurters, or walking wounded." After specifying the presidential qualities that "a grown-up divorced woman" has a right to expect, Donovan goes on to describe the miserable lapses in taste committed by the last six gentlemen who answered the ad.

In Donovan's interpretation, the United States becomes synonymous with the woman of sensibility—WASP, well-educated, obsessed with hygiene, "too intelligent and alive to be happy with loser or numbskull."

As might be expected of a man who once edited *Time,* Donovan has an unparalleled gift for cliché, and his essay fits the standard specifications for the high-minded elegies composed upon the wish for kings. With only slight revisions and amendments the essay could have appeared as an article in *Foreign Affairs,* as a symposium in *Commentary,* as a sermon delivered from the pulpit of the National Cathedral.

The observation that the civilization of the West lacks leadership has by now become so safe and so meaningless that it appears in every speechwriter's lexicon of unassailable truth. No important personage these days makes what his fuglemen advertise as a major statement without alluding at least once to the deterioration of standards and the collapse of authority. Politician and corporate grandee, newspaper columnist and locker-room attendant, cab driver and poet—all agree that only leadership can rescue the West from the pit of anarchy.

Having made this announcement, the speaker usually mentions one of the several stock examples thought to possess the proper sheen of significance. Depending on the mood and the occasion, he refers to the impotence of the United Nations, the decadence of the American political system, the dissolution of the NATO alliance, illiteracy in the schools, the short-term thinking characteristic not only of industry and the credit markets but also of the media. All present nod their heads and know that they have been made wise.

Why, of course, say the portly gentlemen gathered on the terrace or in the ballroom, leadership. By God, it's the leadership that's missing. We can send a man to the moon, but we can't balance the budget or clean up the wreckage in Detroit, and do you know why? For the same reason that the country is being governed by a pack of fools. Because the leadership has gone out of our lives.

The lament for the loss of leadership has an ennobling effect because nobody knows what the word means. Most of the gentlemen who mourn its inexplicable absence wouldn't know a leader if they saw one. If they had the bad luck to come across a leader they would find out that he might demand something from them, and this impertinence would put an abrupt and indignant end to their wish for his return. After all, when Christ showed up in Jerusalem saying the kind of thing that leaders have an awkward habit of saying, his countrymen felt compelled to discourage what seemed to them an overzealous display of leadership.

It never occurs to the editorial writers or the organizers of conferences that leaders practicing leadership might require them to make a moral effort, which, if carried beyond the elementary lessons— like giving up cigarettes and the third whiskey before dinner—might seriously interfere with everybody's habitual round of pleasure and hypocrisy. The gentlemen want the kind of leadership that makes a brave show in the world but doesn't cost much more than the annual dues charged by the Council on Foreign Relations. This is another way of saying that they would prefer, no matter how loud their protests to the contrary, as little leadership as possible.

The disingenuousness of the lament becomes apparent in the context of the surrounding adjectives. The portly gentlemen speak of leadership as if it were an innate quality or skill, vaguely comparable to a talent for playing the piano or solving crossword puzzles. They think of it as a function of personality, an endowment of charisma, a gift akin to a despot's cruelty or a salesman's winning smile. When casting around for exemplary proof of leadership, they mention surgeons, military commanders, and football coaches, sometimes lawyers (if they have had enough experience arranging bribes) or police captains (if they have presided over a precinct known both for the savagery of its residents and the corruption of its elected officials). Probably it is safe to assume that the man most people have in mind for the job would bear a comforting resemblance to Muammar Qaddafi. A man of lighter color and broader education, of course, but still a handsome, authoritarian fellow who plays polo, commands the unswerving loyalty of his troops, and silences the disloyal gabbling of the press.

Although at odds with the pious and deliberately opaque image of a leader that appears in public speeches, this somewhat bleaker portrait adheres more closely to the unspoken wish hidden behind the cloak of the antiphonal murmuring. The sporting and professional analogies establish a false comparison. Leadership consists not in degrees of technique but in traits of character; it requires moral rather than athletic or intellectual effort, and it imposes on both leader and follower the burdens of self-restraint. Edmund Burke put the proposition as follows: "Society cannot exist unless a controlling power upon will and appetite be placed somewhere, and the less of it there is within, the more there must be without." This, of course, is the recognition from which the leadership conferences conveniently turn away.

A similarly deliberate vagueness obscures the discussion of elites and elitism. Although dimly aware of the necessary relation between leadership and an aristocratic principle of government, the people who wonder where all the leaders went dare not say too much about the corollary absence of an elite. In the American conversation the word *elitism* carries with it the connotation of an insult. Elitism stands as a euphemistic surrogate for the forbidden subject of social class. Among people obliged to profess loyalty to the doctrines of egalitarianism, an elitist belongs, by definition, to the party of reaction.

But leadership cannot exist except in alliance with an elite willing to maintain the social forms and institutions that express an ethical ideal—not only of the state, but also of human character. Lacking the support of something like an aristocracy, the man who would be a leader has no choice but to become a demagogue. Nobody likes to concede, at least not in public, this fairly obvious and fundamental fact. In private conversation, perhaps, muttering over drinks at the yacht club, it might be possible to say something about the uses of an elite. But when dictating letters to the *New York Times* or making announcements to the voters and the stockholders, the supposedly responsible voices of the day fall prudently silent.

Only an elitist would blame them for their cowardice, which is perhaps another of the reasons elitism has fallen from favor. Our portly gentlemen belong to an accommodating class, happy to make themselves useful to the wisdom in office. Who wants to argue with the spirit of the times? Better to go along in order to get along.

The egalitarian doctrine that holds as self-evident the freedom and equality of everybody in the room justifies the subsidiary axiom that nobody is better than anybody else. The country was founded upon the premise of boundlessly expanding desire—for goods, for land, for experience, for wealth, for fame. Envisioning a romantic panorama of man at play in the meadows of paradise, of man set free from the constraints of laws and schools, free to constitute himself as his own government, free to declare himself a god, the Americans directed their restless energies toward the gratification of will and appetite. Thomas Jefferson already understood that the Americans would end by defining money as the supreme good because only money could be changed into all the other goods, spiritual as well as temporal. Toward the end of his life, in a letter to a friend, Jefferson remarked: "Money, not morality, is the principle of commercial nations."

Precisely so, but in order to keep their mythology intact, the Americans distinguish between two kinds of elite—the technical and the ornamental. Nearly all Americans honor what they can perceive as honest achievement. They willingly grant the prerogatives of authority to those elites that found their claims on the exhibition of talent, skill, or knowledge so narrowly defined as to fit within the categories of a profession. Doctors and lawyers belong to the technical elite; so do engineers, ballplayers, junior army officers, police detectives, plumbers, research scientists, ship captains, parish priests, and film stars. Nobody quarrels with any of these elitists as long as they confine their remarks to the arenas of their licensed competence. But once they overstep the bounds of their specialties—once the doctor begins to diagnose the health of the human condition, or the lawyer presumes to speak on behalf of the higher consciousness, or the matinee idol presents his views on foreign policy—then the audience grows restive and suspicious. Somewhere in the back of the room somebody raises the familiar American objection, "Oh yeah? Sez who?" Every citizen sets himself up as his own moral entrepreneur. Candidates for political office inevitably campaign against Washington and "the system."

This is what makes it so difficult for the United States to found an elite on an ideal of moral character or civic virtue. Who can prove that the fine phrases consist of anything more substantial than pompous sham? What profits do they earn? The law is what you can get a judge to say is the law, and if he is too dim-witted to know the truth when he sees it, maybe he can be bought for cash. Consider Burke's definition of an English gentleman as a man who "fears God, looks up with awe to kings, with affection to parliaments, with duty to magistrates, with reverence to priests, with respect to nobility." Consider next how well, or how poorly, that definition would fit the gentlemen on the terrace.

Recognizing the national preference for a commercial morality, and recognizing further that sooner or later so cynical a conception of the citizen's obligation to the state could get dangerously out of hand, the American directorate goes to a great deal of trouble and expense to manufacture an ornamental elite.

This is why it costs upwards of $12,000 a year to go to Harvard or Yale. It isn't that those universities offer an education significantly better than the education offered by Michigan State or the University of Alabama, nor is it that they inculcate in their students a love of learning or a habit of rigorous thought. They confer the insignia of wisdom and power. Under the rule of appearances, the Yale or Harvard graduate presumably stands that much closer to the inner chambers of authority. The French aristocracy distinguished itself by the placing of an honorific *de* in front of its names. In the United States the equivalent patents of nobility are derived from a connection with one or more of the relatively small number of institutions that constitute the American peerage—the leading universities, certain New England preparatory schools, various strategic centers, foundations, and research institutes, maybe six or seven clubs, some of the departments of Federal government, and a few of the larger media bureaucracies.

As might be expected in a commercial country, membership in these ornamental elites is sold for money or bestowed as a gift of patronage. The selling of secular offices, analogous to the medieval practice of simony, arouses in the distant public the feeling of envy and resentment; among the important personages, safely gathered inside the walls, the dim awareness of their venal association gives rise to a collective queasiness of conscience. Too many of those present know how dubious are their credentials and how empty their claims to moral authority. They behave like fugitives traveling on false passports, and they suspect that they might have paid too high a price for the prizes of their ambition.

Regrettably, and much to everybody's sorrow or annoyance, this state of affairs cannot be avoided under the present rules. To a larger extent than most people like to admit, the success of the American economy depends on a good deal of sharp dealing that under a strict rendering of moral accounts would be construed as dishonest. The automobile industry makes second-rate cars which it attempts to sell by means of false advertising; consultants peddle expensive gibberish to clients too scared to dismiss an incompetent board of directors; banks feed on the difference between the rate at which they borrow money from the government and the rate at which they lend it to their customers. This may not be particularly surprising or new, but it causes the American elites some embarrassment when it comes to wearing their honorific badges and titles. They fidget in their ceremonial robes and wonder exactly whom the speaker has in mind when he talks about the rewards of integrity and hard work. Machiavelli might have been able to provide a rationale for them, as might Francis Bacon, who comforted his Elizabethan audience with sophisms in praise of the means, "sometimes base," by which men "through indignities come to dignities." But cynicism does not sit well with the American conscience. The American is a sentimental fellow who wants to believe that his words conform to his deeds, that his success deserves the applause of God. To this noble but implausible end he commissions, at vast expense, the symphonies of congratulatory prose delivered in the form of speeches at annual meetings and trade conventions. If only he can make the music loud enough, if only he can drape enough red, white, and blue bunting over the dais, maybe he can shout down the voices of conscience.

No matter how vehemently denied or how imperfectly grasped, it is the knowledge of an inward fraud, of something having gone seriously wrong, that makes it impossible for the Americans to put much faith in the ornamental elites they take so much trouble to fabricate. Like the automobile manufacturers in Detroit, they cannot believe in the worth of their own product. When pressed by circumstances and frightened by the storm of the world, the elites look for their salvation to magical apparitions—to the opportunist and the adventurer, to Henry Kissinger or Zbigniew Brzezinski. The ornaments blow away like straw in the wind, not because of the envy and rage among the people outside the walls, but because the members seated in the club dining room have become sickened by the smell of their own bad faith. The members can no longer distinguish their enemies from their friends. They know what to say to their stockholders, but not to their children.

In order to establish a confident elite—an elite capable of acting in a way consistent with the phrasing of its Fourth of July speeches—the Americans would need to assign a higher value to the unfleshly forms of success that never make the gossip columns. By holding to the maxim that "nice guys finish last" and looking upon the acts of self-denial as the sure marks of a loser, the Americans make of the most deep-rooted human conflict the stuff of musical comedy.

Man wages bitter and incessant war against himself, his nobler elements joined in mortal combat with the old and primitive enemy come against him under the glittering banners of ignorance, superstition, greed, selfishness, and fear. It is this struggle that releases his highest energies, gives rise to his greatest thought, calls forth his most heroic victories. It is in the pit of every man's soul that civilization defends itself against barbarism.

Like the Marxist whom they may profess to despise, Americans change the venue of the argument to the playing fields of politics, and by so doing they seek to make it small. Steadfastly refusing to grant the authority of spirit over external affairs, they conceive of society as a stadium in which the opposing teams—rich and poor, East and West, black and white, capital and labor, North and South— ceaselessly play a game of capture the flag. This conception of the human predicament makes ethics as superfluous as tragedy; it substitutes jogging for thought, slogan for metaphor, novelty for art.

It is the unerring instinct for the trivia that qualifies an American politician as a candidate for office. During the interminable political season nobody ever asks the candidate what Utopia he would build in the wilderness if all his promises could be redeemed and all his social programs changed into the currency of law. What dream of justice does he pursue through the long and exhausting months of photo opportunities, airport press briefings, and noon appearances in suburban shopping malls?

Nobody asks the questions because the answers would tend to obliterate the distinctions, so slight and yet so expensively promoted, between candidates of nominally opposing views. The mediocrity of their collective political imagination leads every candidate to conceive of the American Eden as something very much like one of the Crimean resort hotels constructed by the Soviets as a replica of the worker's paradise or like the dream of the good life sold in Bloomingdale's. The ancillary ideological arguments amount to little more than quibbling about the quality and cost. Were they to be translated into a travel agent's prose, the political attraction of Hotel America might be advertised as follows:

The State. A fanciful name for the hotel management, deserving of respect in the exact degree to which it satisfies the whims of its patrons and meets the public expectation of convenience and style. The guests have no obligation to the state except to pay their bills. This meager definition differs only slightly from the Mafia's designation of itself as *Cosa Nostra* ("our thing").

Freedom. The right to indulge a holiday lust for goods and experience. The guarantee of happiness is included in the price of the room. The management assigns to its guests different classes of license—presidential, ambassador, family, economy, immigrant—but these can be revised upon payment of an appropriate fee.

The Electorate. Understood as the personification of will and appetite. The guests expect a good time. They prefer to leave the making of a moral effort at home, with the wife and kids.

The Laws. Construed not as the permanent, ethical self of the nation but rather as tools with which to gouge from the carcass of the body politic a brigand's share of the feast.

Politics. A Greek word for the printed forms on which the guests can "take a few minutes" to jot down their complaints and suggestions. Every two years the hotel collects these opinions—about the freshness of the orange juice, the enthusiasm of the staff, the placement of the tennis courts—and after submitting the results to the media, the management may decide to hire another wine steward or paint the bedrooms blue.

The Good Life. On sale around the clock in the dining room and the lounge as well as in the international shops located in the mezzanine arcade. The management takes pride in its ability to maintain an Old World atmosphere that reflects a state of being rather than of becoming. The latter condition implies movement, which creates friction, which causes pain, which is unconstitutional.

Among an elite deserving of the name, the emphasis would fall not upon the success of the outward lie—no matter how profitable or fulsomely celebrated in the press—but on the validity of the inward truth. No matter how artful their publicists, most people know when they are lying, and to keep the knowledge at a respectful distance they pay a heavy ransom not only to the suppliers of drugs and pious speeches but also to the makers of sexual and political illusions.

The gift for *trompe l'oeil* has been plain to see in the continuing saga of John Z. DeLorean, late of General Motors, Belfast, and the Beverly Hills celebrity circuit. Some days after he was arrested in Los Angeles on charges of attempting a $60 million cocaine deal, Helen Gurley Brown, editor of *Cosmopolitan,* told reporters that "a hero has been felled." Mrs. Brown can usually be counted on to say something silly, but on this occasion she surpassed the standard of golden inanity that distinguishes her magazine as well as her recent best-selling book, *Having It All.*

Like Mrs. Brown, the media did their best to burnish DeLorean's image with a sheen of glamour. Although they told his story with occasional lapses into ambivalence, not being sure whether to cast him in the role of hero or villain, they tended to favor the straight-forward plot. Ambitious young engineer rises from poverty in Detroit to the citadels of wealth and power; boardroom struggle leads to resignation of heroic General Motors executive who wants to man-ufacture "an ethical car"; maverick entrepreneur goes to Northern Ireland and there, among the ruins of a civil war, sets himself to build a factory; company falls on hard times, and the latter-day Faust, wearing gold chains and California denim, strikes a bargain with Mephistopheles in order to salvage his great and purifying dream. Under somewhat closer textual analysis, the story resolves into a prolonged and relentless cliché. What emerges is the portrait of a man remarkable only for the wolfishness of his appetite. The reader looks in vain for a trace of originality or thought, for a single glimmering of imagination. It is the story of a stomach.

What DeLorean knew was how to consume what he had been told was the best of everything—real estate, shirts, women, hair sprays, airplane tickets, napkins, box seats, publicity, tennis balls. The entire romance of his story consists of nothing but the magical recitation of brand names that grow like sweet fruits in the consumer's garden of paradise. DeLorean drives three Mercedes automobiles. DeLorean wears Gucci shoes and owns an interest in the New York Yankees and the San Diego Chargers. DeLorean eats at LaCôte Basque and Maxim's. DeLorean stays at the Savoy. DeLorean owns a Fifth Avenue apartment, a New Jersey estate, a ranch in San Diego. DeLorean has bank accounts in Switzerland and the Caribbean. DeLorean hangs around with celebrities so expensive they have become collectible.

Against the din of incessant propaganda a man must carry in his mind a nobler conception of himself than the images offered in the show windows of the id. He must feel himself a part of a larger ideal, of a wider self and a longer human tradition than can be contained within a television broadcast. The egalitarian spirit remains suspicious of all things hereditary—of usage, form, custom, habit, iconography, and degree—but it is precisely the business of an elite to preserve this partnership between the living and the dead in order that men might sustain their moral imagination and thus, by imitation, rise to greatness. How else can they acquire what used to be called a sense of honor or conscience?

Anybody who wishes to reflect on the current condition of the American elites has only to compare the character of George Marshall, the secretary of state during the war, and the character of Henry Kissinger. When Marshall retired, having accomplished a good deal more in five years than did Kissinger in all the years of fretful traveling and excited press interviews, he was besieged by New York publishers imploring him to write his memoirs. Marshall declined the assurances of money and celebrity. No, he said in effect, I did nothing other than what was expected of me as a citizen; I see no reason to ransack the State Department files for the sake of gain. Not only did Kissinger take the papers, he employed legal strategems to hold some of them out of the public domain and edited others to coincide with what he thought should be his starring role in the history of nations.

Marshall could not have acted as he did unless he held himself responsible to a code of conduct that kept in check the urgings of will and appetite that he shared, as do all other men, with Kissinger and DeLorean. He was a man in the habit of looking up—toward an ideal that he set for himself as well as for others. Who among the members of the contemporary elites would dare to risk so dangerous and so unprofitable a posture? Let them for a moment take their eyes off the ground, and somebody might steal from them a crumb of advantage.

Anybody who would improve on this state of affairs comes up against odds heavy enough to interest the makers of Hollywood epics. Most men do not have much of an ear for sermons, and the desires raging in the blood rush headlong to whatever wine-dark sea washes the nearest Dionysian coast. Allied with the strength of the instincts, the intellectual bias of the age opposes the very idea of leadership. Of what use to preach the lessons of moderation to people who have proclaimed themselves kings? The egalitarian philosophers hold fast to the illusion that all men have been created not only equal but also free. They fail to notice that men must make their freedom and that the work is both daily and hard. The advertisements for reality offer a more indolent interpretation, encouraging the customers to believe that the rich man is by definition free even if he employs his wealth to satisfy the thousand whims and appetites to which he owes the submission of a slave.

Against the world's weight mankind possesses the greater energy of spirit formed by the mass of small but numberless acts of ordinary human beings going about the extraordinary business of telling the truth. Not the large and mystical expressions of human solidarity, not religion of humanity as defined by the Catholic church or the Ford Foundation, nothing abstract. The negotiations at every dinner table thus become as central to the human story as discussions at Vienna or Versailles; every hospital room bears witness to battles as critical to the human destiny as the battles fought at Gettysburg or Waterloo; every bedroom door opens into a desert or a garden.

The task that confronts the men who would be leaders in any of the democracies is a task of the imagination. It has less to do with politics than it does with metaphor, with the making of new myths that allow men to see their immortality not in their monuments but in their children. Maybe it is an impossible task, like the juggler's dream of the balls standing still in the air, but certainly it is of heroic enough proportion to summon leaders capable of drawing swords from stones.

FOREIGN
PERSPECTIVES

Saul Steinberg, *Passport Photos*

Technology and Freedom in the Soviet Union

S. Frederick Starr

H istory," declared Nikita Khrushchev, "is on our side." With this bold assertion, the premier of the Soviet Union affirmed the official Soviet belief that events within the country and around the world must inevitably unfold according to the prescriptions of Marx and Lenin. Abroad this was thought to mean the collapse of capitalism and the establishment of regimes of the Soviet type. At home it meant the emergence of something called pure communism.

Today this hope has waned, and so has the ideology from which it sprang. It is considered bad form in the Soviet Union to sprinkle one's speech too liberally with quotations from the Communist Founding Fathers—to do so is the sure mark of a middlebrow upbringing. Even the recent Western enthusiasm for the less dogmatic early writings of Marx strikes no response among Soviet intellectuals. In the generation since Khrushchev, old ideals have faded, while new ones have yet to appear. A country that fixed its eyes for half a century on a certain future now sees only question marks.

The Brave New World of NTR

For all the public disillusionment, the Communist party still clings to the old belief that history will eventually confirm Marxist theory. The tenacity of the party on this point provides steady patronage for a small army of official intellectuals, whose task is to revive the old verities by translating them into modern terms. The goal of this group of Marxist theoreticians is to demonstrate that the so-called scientific-technical revolution, universally known in the Soviet Union by its Russian initials, NTR, will eventually transform world society along the lines predicted by Marx. Hundreds of books and articles on the scientific-technical revolution have appeared in the Soviet Union. These works, little studied in the West, contain the official Soviet view of the future. As such, they are a prime source of information on the way official circles in the Soviet Union regard the evolution of human freedom in their country. Will the Soviet system further strengthen the types of control of the individual predicted by George Orwell on the basis of his exposure to the Stalin regime, or will it instead relax those controls in favor of greater individual self-determination?

The NTR literature presents a complex and in some respects extremely diverse picture, but the common thread that unites all these works is a deep conservatism. Most of the participating authors admit that the advent of high technology marks a revolutionary change in human affairs. The information revolution, the modernization of smokestack industries, the integration of the world economy, and the concomitant changes in leadership and management are all seen as affirming existing directions of official policy in the Soviet Union. For most—albeit not all—Soviet theorists studying the effects of high technology on society, the scientific-technical revolution is more scientific and technical than revolutionary.

Soviet notions about the future of virtually every aspect of their society reflect this caution. Nowhere is this more evident than in respect to human rights and freedoms. Without pausing on the dissenting voices, one can speak of two main aspects of the prevailing view of technology as it affects individual freedom. First, the age of high technology will finally establish the negative freedoms from hunger and the like that are enshrined in the Atlantic Charter, the United Nations Charter, and the Soviet Constitution. Second, it will gradually eliminate the social distinctions that now divide Soviet citizens and that give rise to the continued need for coercion by the government. Third, and most important, by abolishing the last vestiges of a society organized around classes, high technology will eliminate the ground on which apologists of bourgeois freedoms are said to feed. In other words, the very ideal of individual rights *against* the state will fade as the conditions that necessitated the existence of a coercive state are removed.

It is impossible either to confirm or to deny the validity of this prognosis. But it is possible to identify areas in which the new technology is already influencing Soviet society and ask whether such changes have a common direction. Will technological advances alter the prevailing patterns of Soviet life? Or will Soviet life as it now exists mold and transform the new technology to its own image?

American observers must first contend with their own biases on these issues. It has long been acknowledged that computer technology affects society in two quite different ways. On the one hand, it affords unprecedented opportunities for centralized coordination and control. Large national and transnational data banks reduce the citizen's realm of privacy and subject him to new forms of public and private scrutiny. On the other hand, computers can exert strong pressures for individualization and decentralization. As desk-top computers infuse new vitality into small enterprises and enable individuals to make well-informed decisions on important questions that affect their lives, they enlarge the individual's sphere of choice. Romantic populists take this possibility to its logical extreme, pointing to a day when computers will make possible new forms of consensual government, under which national plebiscites could be conducted on virtually any public issue within minutes.

While the two tendencies are in contention, recent developments in the United States encourage those who stress the freedom-enhancing aspects of the new technology. These optimists see it as an alternative to coercive bureaucracy and the high road to public efficiencies that will release time for the individual to develop his private realm. Such people consider the computer to be yet one more phase in a series of technological changes that have helped emancipate the individual from social coercion. Like Soviet theoreticians of the scientific-technical revolution, they are convinced that the new technologies constitute the apotheosis of the basic existing values of their society. But while the Soviet theorists see computers as leading to the creation of the perfect collective society, their American counterparts look to computers to fulfill the promise of the Declaration of Independence and the Bill of Rights.

Let us allow the possibility that both perspectives are to some degree valid. The clear implication would be that, Marx notwithstanding, political and social systems transform the economy at least as much as they are affected by them. Technology, according to this view, is neutral; its tendency toward freedom or coercion is derived more from its human context than from its inner nature. Its further advance in the Soviet Union is not likely to weaken the extensive controls over the lives of individuals that is the living heritage of Stalin's quarter century in power.

The cornerstone of this conclusion is obviously the proposition that societies transform technologies. If this assertion is valid, then one should expect to find in Russian history other instances of the prevailing statist, centralized, and authoritarian system molding new technologies in its own image. As it happens, Russian history is full of dramatic instances of precisely this phenomenon. The introduction of printing from movable type during the 1550s and the advent of vehicles powered by the internal-combustion engine in the twentieth century are particularly vivid cases in point.

Technological Revolutions in Russia's Past
A generation ago, Marshall McLuhan argued that the very medium of printing democratized the flow of information. Thanks to Gutenberg's invention, new ideas could be disseminated freely from hundreds of points. As a result, each individual could command the information necessary to form his own judgments, without the mediation of priests or princes.

In Russia, by contrast, the introduction of Gutenberg's epochal invention had precisely the opposite result. For a quarter of a millenium, printing in Russia was monopolized by the state. Virtually all presses were built and maintained by the government, and most were centralized in Moscow. They were used to disseminate standardized copies of such routine matters as laws and decrees. The function of printing in Muscovy and in early Imperial Russia was to centralize control over information so as to render the tsarist state more homogeneous and efficient.

The new ideas of the era were, for the most part, disseminated through the old-fashioned media of manuscripts and wood-block printing. Even the introduction of steam presses, which heralded the rise of a free and pluralistic public press in the West, served opposite ends in Imperial Russia. Whereas the first steam press in England belonged to the *Times* of London, the first such press in Russia was established by the Ministry of Internal Affairs.

To be sure, a more pluralistic system of printing eventually developed in nineteenth-century Russia. But as this occurred, the state moved rapidly to subject that system to official control through censorship. As a result, many of the finest achievements of Russian letters continued to be disseminated first through the old-fashioned medium of manuscript. And with notable exceptions, the most modern printing technology continued to be devoted mainly to official printing and to the publication of carefully censored works for the mass market.

The heart of the tsarist Russian political system was the primacy of the state. This principle suffused virtually every aspect of Russian life, creating a system based on duties rather than on rights. All the principal technological changes of the nineteenth and early twentieth centuries passed through this filter. Railroads, steam engines, telegraphy, and the internal-combustion engine all had fates in Russia similar to that of printing. In each instance the new technology was molded by the prevailing political culture so as to strengthen rather than change it.

Automobiles and Freedom

Until recent years, internal-combustion engines were used mainly in state-owned trucks, not in individually owned automobiles. Will mass production of Soviet-made Fiats herald a change in this fundamental pattern? Will private automobiles contribute to the creation of new forms of individual freedom that were unknown a generation ago?

There is much to support such a view, for the private automobile is undeniably an instrument of individual freedom. In the Soviet Union it enables citizens to liberate themselves from the regimentation of mass public transport and of employer-sponsored group vacations. Instead of traveling as a *kollektiv* to some overorganized company spa, Soviet citizens who own automobiles are able to roam the back roads of their country alone or in intimate groups of family or friends. The automobile thus opens new horizons. It enhances freedom from state tutelage and also from the pervasive claims of society.

Automobiles also open vast fields to private entrepreneurship in the Soviet Union. Repair work, spare parts, accessories, and used-car sales all provide employment for thousands of hustling individual businessmen. Differences in the costs of such goods and services in various cities have enabled enterprising Soviet citizens to organize their private businesses on a national scale. Because automobiles are distributed without regard for regional variations in wealth, weather, and road conditions, for example, people in the poor and underdeveloped North regularly earn money by selling their cars to "dealers" in the prosperous and automobile-hungry Baltic.

Yet for every individualizing aspect of the automobile revolution, there is a collective counterweight. Brochures published by the Soviet automobile club sing the praises of vacationing by convoy in automobiles. Hostels, hotels, and motels give large price breaks to groups, thus imposing a tax on the individual traveler. Collective control over vehicle maintenance extends to the most minute details: automobile owners who drive dirty cars are subject to fines! Checkpoints along major highways maintain the discipline of the internal passport system.

By means of such measures, the individualizing tendency of privately owned automobiles in the Soviet Union has been contained. This example reminds us that in studying the effects of high technology in the Soviet Union it is not enough merely to plot the steps by which a given innovation advances. The key variable is, rather, the policies by which the state seeks to confine the effects of that technology within acceptable bounds.

Computer Communism
It is in this context that the so-called computer revolution must be approached. Between the Soviet computer industry centered in Byelorussia and the closely related Eastern European computer industry headquartered in the German Democratic Republic, the Soviet Union today has access to a wide variety of computer hardware and software. Imports from Western Europe and Asia further expand the range of equipment available. Even though the gap between it an the West has recently broadened after a period of narrowing, the Soviet Union still has computer resources that are adequate to the fulfillment of many important social needs, both military and civilian.

Major industries, such as the production of hydroelectric turbines, are heavily computerized. In the most advanced sectors of the economy, computers not only guide the process of manufacturing but also monitor purchasing, inventory control, and bookkeeping.

Coordinating the entire Soviet economy is the State Planning Agency, which commands an extensive computer system, not only at its Moscow headquarters but also at its fifteen republican branches, all of which are linked closely with the center.

Economists in the larger government enterprises regularly construct computerized models as a means of assessing policy options, much as their counterparts do in the West. The State Bank enters domestic and international transfers into its centralized computer network, which it then uses to plot trends and to evaluate progress toward centrally established goals. Still more extensive is the application of computers to military needs and intelligence services.

Conspicuously absent from the list of Soviet enterprises that employ computers are those engaged in the so-called second economy. This twilight realm of semilegal private enterprise now accounts for approximately 25 percent of the Soviet gross national product. Its importance to the Soviet economy as a whole has received much attention from Western Kremlinologists. Nonetheless, the second economy remains largely untouched by the computer revolution. If this continues, the unofficial private sector will gradually become a technological backwater, no matter how productive it may be, with obvious consequences for the future of individual freedoms in the Soviet Union.

In contrast to the West, computers in the Soviet Union remain overwhelmingly a producer good. Manufactured or imported by and for the state, computers have been used to render the existing governmental and administrative system more effective, not to change it. Far from broadening the sphere of pluralism and private choice, computers have emerged as the last best hope for making the old command economy work.

Given this pessimistic outlook, one cannot help but wonder whether there are not other forces besides the scientific-technical revolution that might ameliorate the effects of the command economy on individual freedoms in the Soviet Union.

The standard line of argument on this point among Western specialists on the Soviet Union is frankly apocalyptic. It holds that the centralized system inherited from Stalin does not work and sooner or later must necessarily be replaced by a system that permits more individual initiative. But this notion, trumpeted by leading Kremlinologists, students of the Soviet economy, and newspaper columnists, contains a double fallacy. On the one hand, it is based on the assumption that the Soviet economy is in more of a shambles than it actually is. With GNP in the first six months of 1983 running 5 percent ahead of the preceding year, most Soviet planners see little cause for despair. True, many goods are in pitifully short supply, and those that are produced are long on quantity and short on quality. It is also true that the productivity of labor has been declining at a time when the Soviet Union faces a labor shortage of serious proportions. Yet the quality of life of most Soviet citizens has improved during the past generation. Even if the salaries earned by members of the intelligentsia have stagnated, factory wages have increased steadily, at least until recently.

All this has occurred even at a time when the country has chosen to invest staggering sums in the military. Such considerations leave most Soviet leaders convinced that, while certain changes are desirable and necessary, the system itself is fundamentally sound.

Even if this were not true, the universal desire to preserve domestic tranquillity, the absence of a shared sense of economic crisis, and the pride that comes from the knowledge that one's country is a superpower would all serve to protect the status quo. The present system is far more likely to erode than to explode. Barring a sudden and dramatic downturn, the Soviet Union will continue to muddle through.

Assuming, then, that the new technology need not enhance civic freedoms in the Soviet Union, are there no other forces for positive change in that country? Is Russia on the eve of 1984 fated to pursue a course of development that will push it further toward the realization of George Orwell's nightmarish vision?

The Conformist Peasant
The point on which any answer to this question must turn is Russia's distinctive political culture. Based upon duties to the state rather than on rights against it and on the primacy of the collective over the individual, Russia's inherited political culture reinforces the Communist system that shapes daily life. The roots of this political culture lie deep in the Russian past, specifically in the world of the Russian peasantry.

Rural Russia was never dotted with private farmsteads of the kind that produced hardy individualists in New England. Instead, groups of families lived together in village communes and worked the land collectively. Not only did they labor as groups but they also paid taxes and served in the army as groups. Serfdom, an authoritarian though remote state, and a form of Orthodox Christianity that exalted the collective over the individual undergirded communal life. Communal values were also reinforced by the patriarchal family and by the fact that young peasants passed directly from childhood to adulthood, with no period of youthful independence and experimentation in between.

When members of the village commune had to reach decisions about planting or harvesting, they did so through a Quaker-type consensus. When they could not agree, as often happened, the elder simply laid down the law. His decision was final, and anyone who disagreed with it was expected to fall into line. Conformity and submissiveness were part of the system, as was endurance, for the life of a Russian peasant was hard.

Collective agriculture shaped peasant Russia's attitude toward deviance. Dissident elements were suppressed, or, if they could not be suppressed, banished from the village. Solzhenitsyn's exile to Vermont is a recent example of this age-old practice at the national level. The goal of such actions was to defend the collective society from dangers, real or imagined. Besides protecting itself against internal rebels and dissidents, the communal village had also to be on guard against external threats from landlords, the government, and do-gooders. All were treated with the same combination of suspicion, cunning, and outright hostility.

Formed in the course of many centuries, the outlook of the Russian peasant handily survived the various agrarian reforms introduced both before and after the Bolshevik Revolution of 1917. Stalin's forceful reorganization of agriculture during the years 1929–32 left millions of peasants dead but in the end buttressed the peasant mentality by once more binding the rural populace to collective farms.

City Air Makes Men Free
Against this background, the recent tide of persons choosing to move out of collectivized agriculture and into the cities is seen to be as important as any change that has taken place in Russia during the past millenium. As late as 1950, two thirds of the Soviet population was rural. Today, the proportion has dropped to less than one third and it is still falling. More than a million and a half people flee the farm each year and have been doing so for a decade.

Peasant Russia is dying, and the numbers tell only part of the story. An overwhelming proportion of those who have immigrated to the cities are the young, the talented, and the ambitious. Rural Russia today is inhabited by the elderly and the also-rans, along with a small but growing group of Soviet-style agroindustrialists who have as little sympathy for traditional peasant life as do those who left for the cities.

Soviet leaders long hoped that village conformity and discipline could somehow be retained in the cities. But cities impose their own values, and these are rapidly transforming the outlook of Russia's former peasants. Urban life in the Soviet Union, no less than elsewhere, means large apartment complexes, individualization, fragmentation. It also means education, access to expanded sources of information, and relative freedom from parental authority and tradition. It means the freedom to change jobs, a freedom exercised by a fifth of the labor force each year. Instead of endurance and suffering, it means the hope of gratifying one's ambitions and desires. Finally, it means youth and the entire culture of the young that characterizes the modern world from Seattle to Budapest to Tokyo.

It would be naive to think that the mentality of peasant Russia will vanish overnight. Until recently, it was still being passed on to urban children by peasant grandmothers brought in from the countryside to look after the children of upwardly mobile working parents. But the move to the city is exposing millions of younger Soviet citizens to a world in which people jockey for power in their offices, pore over reports translated from foreign journals, make deals to obtain Scandinavian furniture for their living rooms, and pull strings to enroll their children in selective schools.

This does not mean that the Russian people are about to "converge" with those of the United States and other Western industrial countries. The Communist economic system and authoritarian political regime remain fundamentally different from ours. Yet the psychological distance between the peoples of the Soviet Union and of Western Europe and the United States is definitely narrowing, as the Russian people shift to an urban and more Western way of life. The aspirations and worries of the new Soviet urbanites are far closer to those of their counterparts in the United States than to those of anyone in their own rural past. One can scarcely imagine a bearded Russian peasant spicing his speech with talk of "best-sellers," "weekends," "hyping," and "communication gaps." Yet these issues are all debated in Soviet newspapers and magazines today.

These new social conditions are the essential preconditions for any future changes in the status of human rights and freedoms in the Soviet Union. Their effect is gradual but pervasive. They form what Alexis de Tocqueville called "a kind of intellectual atmosphere in which both governed and governors move and from which they draw the principles of their conduct, often without realizing it."

The Rise of Solo Personalities
Is there evidence that this new intellectual atmosphere is actually coming into being? The majority of the population is less than thirty years old, and the fact that members of this group have lived their entire lives in the post-Stalin era sets them apart from their elders. The measure of the difference this makes can be found in the sheer extent of the intergenerational conflict, which makes itself felt in virtually every Soviet family today. Soviet families have changed almost beyond recognition. Differences in education between fathers and children erode the authority of the traditional paterfamilias. Ripped from their rural environment and thrown into impersonal apartment blocks, individual members of the family are less subject to collective controls, whether from other members of their families or from society

at large. The rise of urban crime is an accurate barometer of the breakdown of traditional collective discipline.

With less ready recourse to control by the collective group, conflicts among individuals are more likely to be adjudicated by impersonal legal processes than ever before. The reliance upon objective legal norms as opposed to dictates of the village collectivity is itself an important expansion of the individual's realm of freedom.

Nor do the clichés of collectivist ideology provide strong values on which individual Soviet citizens base their actions. The widespread search for new principles of conduct finds expression in many forms, among them the shift in student interest from the natural sciences to the humanities. Surveys of high school graduates in the Soviet Union, for example, reveal a rapid rise in the prestige of the profession of writer.

The breakdown of old value systems has led some Soviet young people to take refuge in a native redneck patriotism. Still larger numbers of Soviet youths find their psychological home in the emerging popular culture, which is guided overwhelmingly by creative people whose lives are little subject to state guidance. The basic values of Soviet popular culture today are influenced far more strongly by independent-minded and youthful Soviet trend setters or by their counterparts in the United States, England, or France, than by any official organs within the Soviet Union itself.

Prospects for the Future
The best evidence of the growth of a freer mentality in the Soviet Union is the way minor technologies are being exploited by the younger generation for its own private ends. The former government monopoly of information is far from total today. Where once there were cable radios tuned to the one official station, there are now millions of privately owned short-wave receivers. The ephemera of independent speech and song can now be rendered permanent and transmittable, thanks to the availability of small tape recorders. Even the state monopoly of photoduplicating machines has been weakened, as the adventurous make use of the proliferating copying machines in their offices and institutes. Without denying the severity of the many controls that still exist, one can only be struck by the multiplication of private modes of communication, thanks to the new technologies.

To be sure, the Soviet state maintains its former instruments of control and is even strengthening them, as the KGB and other security agencies acquire their own advanced computer systems. Nonetheless, it has become gradually more difficult for state agencies to exercise the full measure of control over the lives of individual citizens that they are empowered to exert under Soviet law.

In the face of the well-publicized violations of human rights in the Soviet Union, this claim may seem dubious, if not false. Nonetheless, three separate considerations give grounds for thinking that self-restraint by the Soviet state in the treatment of its own population will increase rather than decrease in the coming years.

In the old days, the populations of rural villages could be easily monitored by party representatives stationed there. Today, urban dwellers live an average of one hour and thirty minutes from their places of work and must therefore spend a large part of each day moving about in the city on their own, beyond the reach of authorities. As they free themselves from age-old habits of subordination to the collective, they gain experience in making decisions about their lives and in maximizing their own advantages in a complex world. The number of people in this unfamiliar state has become so great that it is beyond the powers of even the huge Soviet security apparatus to keep track of every individual's actions.

Second, the Soviet Union has an urgent need to motivate skilled persons to work. Neither simple exhortation nor the threat of force serves this end today. While it might be argued that the productivity of Soviet workers will not increase until basic structural and administrative reforms are introduced, it is still evident that lockstep regimentation is self-defeating. As the need for greater productivity grows yet more pressing, as it is bound to do, the necessity of accommodating the actual desires of the urban populace will become unavoidable.

Third, for the Soviet government to exert strict controls over the population, it must have resolute organizations and individuals to carry out its policies. In the past, this task has fallen to the Communist party and to the Committee on State Security (KGB). The effectiveness of both bodies has depended on their status as quasi-military organizations whose esprit de corps and self-discipline set their members off sharply from the population at large. Today, both groups, especially the party, are beset by erosive tendencies.

The Communist party of the Soviet Union is a small and highly disciplined elite, encompassing scarcely 5 percent of the population. It is also an elite that places severe and constant demands upon its members, particularly those who serve as professionals on its staff. At the same time, party membership has been virtually a requirement among the highly educated. Fully two thirds of college-educated urban males of working age hold party cards.

Nevertheless, recent developments have presented the party with a dilemma. Urbanization has been accompanied by a vast increase in higher education. If the party is to maintain its presence among the educated classes, it must increase its membership in proportion to the growth in the number of college graduates. Yet if it pursues this course, the Communist party will become a sprawling and loose organization, incapable of exerting control over society as a disciplined elite. It can maintain its self-discipline and rigor by staying at the present size, but if it follows this course, it will isolate itself from most Soviet citizens with higher education. Either alternative will erode the ability of the party to control Soviet society.

The KGB, too, functions in an environment quite different from that which existed only a generation ago. It was easy to control a small educated class whose members rarely went abroad. Since members of the technical elite had only recently emerged from modest backgrounds, they were subservient and willing to accept harassing controls as the price of their high status. Today, the growing number and sophistication of the technological elite are changing this situation. The rising second generation intellectuals are more self-confident, more venturesome intellectually, and less tolerant of primitive controls over them. Nor can members of the burgeoning intellectual class be excluded from privileged information. The very complexity of the economy and the first steps toward an information society require that millions of persons be entrusted with what was formerly considered sensitive or secret information. Each of these developments renders the task of the KGB more difficult. The recent and unprecedented promotion of men with KGB backgrounds to important positions within the party signifies a growing sense of urgency that surrounds the issue of public regimentation in the Soviet Union.

Goodbye to Nineteen Eighty-Four?

The various factors cited here are grounds for thinking that the sphere of individual freedom in the Soviet Union might expand in the future. Advanced industrial societies have no choice but to devolve initiative on more of their members. To survive, they must gain the support of ever more sophisticated populations. This is all but impossible without a degree of freedom.

Such social changes are the direct consequence of the introduction of advanced technology into Soviet life. As mentioned earlier, however, that technology itself can readily be harnessed to the service of a command economy based on duties rather than rights. Computers can make central control less obtrusive, but they can also serve the cause of centralization. To the extent that computers reinforce centralization in the Soviet Union, tensions will arise between the emerging society, with its large numbers of highly educated members, and the economy that supports it.

Such tensions are by no means unique to the Soviet Union; they are certainly evident in our own society. The way they are resolved in the Soviet Union will define the degree of freedom that prevails in that country. Many forces, including the internal environment, will affect the outcome. If conditions abroad fan official Soviet insecurities, the government will have additional grounds for repressing the natural tendencies of the emerging society. A more tranquil international order, by contrast, will undermine the main justification for the maintenance of old-fashioned controls.

The government of the Soviet Union has declared itself in favor or world peace and unimpeded contacts among peoples. Paradoxically, these will do most to undermine the vestiges of Stalinism and to set free the natural tendencies of social and intellectual evolution in the Soviet Union. International tranquillity and contact across borders, rather than high technology per se, will determine the future of individual freedom in the Soviet Union.

High Technology and Human Freedom: French Perspectives

GERMAINE BRÉE

oday in France, intellectual circles and common people alike, but for different reasons, remain skeptical as to the meaning and human values of contemporary technics." Jean-Claude Beaune's remark appeared in a succinct, brilliant overview of the "philosophy of technology in France in the twentieth century"—Beaune's field of investigation. Today, Beaune concludes, French thinking about technology is a "disorganized whole of disparate reflections," a situation that hardly seems peculiarly French. Actually, our technicized age—with its rapid and vast communications systems, its multiplication of symposia, colloquia, international research teams—makes it rather difficult to sort out national reactions from more generally pervasive ones.

The French, on the whole, have accepted without much trouble the technicized society of our time. It is from French philosophers, sociologists, and historians of science that warning notes have recently been sounded. Daniel Cérézuelle, of the University of Grenoble, recently took French intellectuals to task, whether they were politically rightist or leftist, for failure to confront the questions raised by the accelerating effects of technology on the French environment—apparent in the spiraling bureaucracy, in the so-called efficiency that does not save time but accelerates the tempo of work, and in an accumulation of data and loss of perspective.

One subject of peculiarly French concern is the effect of a technicized society on traditional French culture. That concern, linked to the question of the new technology, was expressed in a recent article entitled "The Beaubourg effect," an article in which the term "high tech," otherwise seldom employed, was used. The term there refers not to the design of a cultural center, but to the attitudes of the human beings who work inside it. "Standing and on the move, the staff affects a laid-back, flexible style; very high-tech, very adapted to the structure of the modern space. But seated in their cubicles—which are hardly cubicles at all—they strain to spin for themselves an artificial bubble."

Thus does Jean Baudrillart, a conservative sociologist, claim that the staff of Beaubourg is working in a state of unconscious psychological tension generated by a conflict between the conditions of work in the modern, "polyvalent" open space, and the traditional private office space. This is the starting point for his analysis of the uneasy coexistance in France and in French sensibilities of two incompatible cultures. The conflict is not the unusual one between culture and counterculture. The adversaries here are, he asserts, a new, still uncertain mass culture of "simulation, fascination and acceleration" and a disappearing, almost defunct culture of "secrecy, seduction, initiation and symbolic exchange ... highly ritualized and restrained, a culture of meaning and aesthetic sensibility." "Simulation, fascination and acceleration" point to the media, and Beaudrillart's position recalls that of McLuhan. It was in McLuhan's terms that the same misgivings recently surfaced in France, opposing the children of Gutenberg to the children of the media.

For nearly a hundred years, French academics have exercised a quasi monopoly as the shapers and distributors of knowledge and cultural values. French culture has been transmitted primarily through a hierarchical and centralized education system. A new configuration has now appeared, a rival system of cultural diffusion. The question was eloquently discussed in 1981 in a special issue of the *Nouvel Observateur*, a widely read weekly magazine. The argument put forward by the academics was that the media curtailed the freedom of thought of individual minds by propagating a single, one-dimensional representation of reality. The complex insights elaborated in the past were trivialized or lost in the new process of communication. This partial and limited attack on the technicizing of communications expressed a resurgence of traditional French attitudes

But at the turn of the century, in France at least, there was no cultural dichotomy. The dawn of the new—sometimes called the second—technological age was hailed by poets and painters who shared excitement. The Eiffel Tower was its emblem; the large transoceanic steamers, the fast transcontinental trains, the cables like "towers of Babel" transmitting a myriad of voices across space, the efficiency and beauty of the powerful new machines, the first airplanes, were motifs to be celebrated: futurists, advocates of *l'esprit nouveau*, orphists, unanimists all praised the freedoms and power newly available to men. Jules Romains, a young novelist, playwright, and poet soon to become prominent, offered his own visionary key to the future: a new psychology, a new ethics, and a dynamic new code of action. "We were the children of the large modern city," he wrote, "and I might say the first it recognized and fully satisfied. The large city breathed around and through us. ... The rumors of its densely populated sections, the vibrations of its faraway suburbs opening out like a fan were the constant music that accompanied our lives."

As a child of the modern world, Romains left behind the romantic concern with the self and its nostalgia for the past, its disdain for "scientific," positivist modes of thought. Modern man, he claimed, must become aware of the new dimensions and powers available to him as an integral member of a community, itself an emitter of psychic energy enhanced by new technologies. Any action, however simple— just walking down a city street, for instance—involved individuals, unbeknown to them, in networks of human intercommunications ever more extensive, ever more powerful: street, city, nation, and, looming over the horizon, soon to become *united,* Europe. Beyond it, in the future, a world community. Émile Durkheim's new socioanthropological studies had sparked young Romain's imagination. Inspired by Durkheim's analysis of group emotions, rituals, and the religious sense they give individuals of a communal participation in the sacred, he even wrote a *petit traité d'autodéification,* which he later repudiated. But he did not abandon his belief that modern technology provided new, freer conditions of life, liberating the working classes, producing new social types.

This mood of exhilaration was widespread. We sense, wrote another young man, Jacques Rivière, summing it all up, that something in our world is changing, "veering like a great steamer which has slipped its moorings and is gliding out to sea." The date was 1913. So strong was the belief—without irony—in a "brave new world," that for some even World War I was discounted as a throwback to the past.

The transfer of technological models or systems to the creation of new art forms, literary forms among them, is one of the main dynamic trends of twentieth-century art, especially French modernist and postmodernist experimental art. At first, this optimism with regard to technology was unconnected to the sociophilosophical attitude so eloquently attacked by Daniel Cerezuelle—a belief in the "good old verities of rationalistic progressivism inherited from Comte, Hegel, and Marx." From the French Revolution on, these had reinforced the underlying humanistic philosophy of progress of the French Republic. In that context, technological developments in and of themselves posed no problems: the problems rested with the use people made of them and raised primarily ethical and social questions. Young French intellectuals were still attuned to the spirit of the Enlightenment, whose credo Gibbon had succinctly expressed: "Every age in the world has increased—and will increase—the real wealth, the happiness, the knowledge and perhaps the virtue of the human race."

Few people in the 1950s, by contrast, would have unhesitatingly accepted the dictum. But the wish to understand and explore the new reaches of human awareness was all the stronger. Thence, in the

1950s, the more austere, more reflexive, urge to create art forms—especially literary forms—to reflect the realities of the present adequately.

Let me give two examples I have selected from the experimental novels of the 1950s. Michael Butor, outstanding among the self-proclaimed "new writers," wished to open his reader's mind and eyes to the power and beauty of Niagara Falls. He considered the falls one of the wonders of our world. For his purpose he created a new literary form, a "stereophonic study," he called it. He entitled his work *6,810,000 Liters a Second.* What he wanted to convey was the sheer power of the sonorous continuum into which all visitors to Niagara are plunged. He worked out a methodical, elaborate, carefully calculated, almost computerized interweaving of voices—the voices of tourists, families, couples, guidebooks, and travel agencies—against the background of Chateaubriand's late eighteenth-century resonant, rhythmical description of the falls. Cycles of night and day, changing seasons, are all drawn into place within the thematic and sonorous whole. The deployment of the poem thus approximates a musical score. It was produced in a new Grenoble theater, itself a high-tech structure, using all the possibilities of a multitrack tape. Each person in the audience, by means of a stereophonic set; could manipulate the blending of the voices at will.

Butor, in this experimental work, was using the new media to double purpose: to reaffirm the written text as voice, itself a continuum of sound and rhythm flowing through the textual surface of the writing. Essentially, perhaps, he hoped to expand and enrich the cultural awareness of his audience by means of an experience lived through all the senses and in a kind of collaboration with writer and media. This may be the kind of activity Marcel Duchamp dreamed of when he spoke of a potential art of "musical sculpture, sounds lasting and leaving from different places and forming a sounding sculpture."

In somewhat the same mood, though with more dubious success, Philippe Sollers has experimented with new forms of writing. In the genesis of his *Paradis*, a taped, then printed text, unpunctuated, unparagraphed, Telstar played a double role as a metaphor and micromodel. *Paradis* catches within its verbal flow, as Telstar might, the medley of discourses alive at any one moment in world and minds, weaving them into a kind of seamless web to be transmitted outward, caught at random perhaps, and partially transmitted from receptor to receptor, through space and time. Sollers wants to convey a notion of literature as a polycentric, polyphonic collective expression of the expanding modern consciousness located on our small globe, rolling space in its "galaxy of galaxies."

What is apparent is that some French artists and writers have deliberately carried on a sustained dialogue, if not a rivalry, with the new media. They have experimented with the properties that language acquires when it interacts with the media, and they have been sensitive, too, to the multilayered, oral-aural reality in which the media immerse us. Theirs is a sometimes esoteric but always valuable attempt to go against the set patterns of literary discourse imposed by the printed page.

Since the 1960s a brilliant group of thinkers—historians and philosophers of science and culture—have also been engaged in an exploration of the assumptions underlying Western culture, specifically French culture. They have been working with tenacity at what Jean Beaune calls a *déblocage-mental*—that is, an attempt to go beyond the taken-for-granted mind-sets that blind individuals and, more dangerously, collectivities, to the emerging patterns of new forces at work in the dynamic processes that are shaping our world. The names of Edgar Morin, Michel Foucault, Michel Serres, and the much maligned Jacques Derrida, are beginning to be well known outside France. One of the questions they have raised concerns the scientists' belief in the ethical and political autonomy of their work, in its unquestioned dedication to the good of humanity. They point to the close connection between scientific research and the industrial and political structures of our society.

In this context, one of the first studies to question the cultural and political implications of modern technology, and to call for a social philosophy that would respond to them, was Jacques Ellul's *La Technique ou l'enjeu du siècle (1954)*. Published more than thirty years ago, it was translated ten years later as *The Technological Society*. The translated title missed the main thrust of his topic—that is, a meditation on what is at stake for our future in the technological revolution. Ellul's argument cannot be summarized easily; but it does mark the passage in France from the ideological concerns of the midcentury to a concern with the uncontrolled spread of monolithic computer systems and their potential for exercising invisible forms of cultural pressure. Ellul sees the proliferation of computers as tending to produce "continuously improved means to carelessly defined ends." He points to the accumulating pressures indirectly brought to bear on our societies to develop the needs engendered by the technology itself, so that social values come to be formulated according to the needs of technology, not of the human community it is supposed to serve. Ellul concludes that the dominance of technique tends insidiously to subvert democracy, making the state an instrument of technology itself, progressively eliminating political debate in the name of maximum efficiency. He does not question the power of scientific research to bring valid answers to those problems it defines. But he questions the processes by which the problems are selected and defined. Even more threatening, according to Ellul, is the chaos that is visibly disrupting our society as a runaway technology launches into circulation anything it has the means to create, if it can be broadly marketed.

Indeed, for the modest onlooker today, chaos rather than a dictatorial Big Brother seems to be engulfing large parts of the world. Ellul, like Orwell, sees a future fraught with danger. Will a new type of technological man emerge, less ruthless than industrial man in the use of the power he wields? Or will technological man show the same drive for personal power and profit as his predecessor? Ellul suggests that we create informed groups of citizens, operating outside the state structure and working to choose among the many technological possibilities those that are compatible with our ethical and cultural values. On this effort, Ellul suggests, the road to the future rests, and he is far from optimistic as to our ability to make it.

There is one road, however, that casts its shadow on any view of the future: the widely publicized world buildup of high-tech weaponry. The French, though less articulate on the topic than some other groups—perhaps because they are more thoroughly trapped in the ambiguities of the international power struggle—find it as difficult as we do to live with nuclear weapons. Under that shadow the future appears more as a dead end than as an open road. Imagination, indispensable in the shaping of new vistas, comes to a halt.

"The faculty of turning away one's eyes as one approaches a chasm," Henry Adams remarked, "is not unknown." Perhaps he had in mind the disturbing figure from the Tarot cards known as *Le Mat,* or the Fool. A figure on the move, he walks along the edge of a precipice, a dog yapping at his heels, his pack flung over his shoulder, his head vacuously turned in the opposite direction. *Le Mat* seems to me emblematic of the uneasy sense the French—like many of us— have of the dangerous road down which we are walking. Thence the growing skepticism noted by Jean-Claude Beaune regarding the miracles of technology abundantly put to use for our profit and pleasure. Thence the effort of the philosophers and historians I have mentioned to define the connecting link between the new objects around us and the hidden drive that produced them and shapes our future.

Like the intellectuals and the artists, the common people in France feel that they are in transit within a strangely shifting landscape in a world of dreams and nightmares. In the meantime they live as best they can in the present, between wonder and potential horror. In the words of Jean Rousselot:

> Between two landings of pilotless planes
> Between two weekends in Venus
> Between two hecatombs.

The Search for Absolutes: Iran

MARY CATHERINE BATESON

Turning and turning in the widening gyre
The falcon cannot hear the falconer;
Things fall apart; the centre cannot hold;
Mere anarchy is loosed upon the world,
The blood-dimmed tide is loosed, and everywhere
The ceremony of innocence is drowned;
The best lack all conviction, while the worst
Are full of passionate intensity.
Surely some revelation is at hand;
Surely the Second Coming is at hand.
—William Butler Yeats
"The Second Coming"

In the winter of 1978, in the town of Bābol Sar on the Caspian coast of Iran, I sat in a family's living room where, evening after evening, the Imam Khomeini made mystical appearances. The head of the household was an engineer who has since been shot in an incident of local fighting. The family were Muslims, although the wife was more inclined toward the Sufi tradition than toward orthodoxy, and the husband enjoyed a glass of vodka in the evening. But they opposed the Shah's government, which they regarded as a tyrannical usurpation, and both had relatives who had been exiled or imprisoned. The family sat in awe, hour after hour, watching the apparitions. The mechanism was not mysterious to anyone present, but this made it no less magical: in the living-room wall there was a large window of frosted glass, which had been smeared during its installation. These smears, like dark eye sockets, combined with the aureole of a street lamp outside to create the vivid impression of Khomeini's face under his huge turban.

This family, like many others, committed itself in the face of great risks to the support of the revolutionary movement, believing that it would bring the country from tyranny to freedom and finding in the enigmatic figure of Khomeini an expression of values they held deeply. We have become so familiar with mass demonstrations of hostility toward the shah, this country, or the hostages, that it was possible to believe that the revolution was motivated by negative emotions. Yet the more pervasive experience in Iran was one of hope and dedication, the feeling in the streets one of euphoria.

There are many similarities between the despotic governments of prerevolutionary and postrevolutionary Iran. The main difference is that the Islamic republican government came into being through the mobilization of a set of feelings that were religious or parareligious. It still depends for much of its authority on a more intimate affirmation than was ever exacted by the shah. The face seen on that window in Bābol Sar symbolized both omnipotence and omniscience; it was a face with piercing eyes, able to see directly into the heart. The mixture of love and awe—the numinous sense that surrounds that face, echoing centuries of hope and longing and martyrdom—has proved to be immensely powerful.

It is important to see the continuities between the forms of social control in prerevolutionary and postrevolutionary Iran. For certain groups, however, the new tyranny is very much worse than the old. It has created a set of conditions that make life more onerous even for groups for which the euphoria of revolution still continues.

The principal continuities are so obvious that they almost invite satire. Most striking is the centrality of the figure of Khomeini as a replacement of the shah. Many analysts are skeptical whether *any* form of government that does not provide a symbolic central figure can survive long in Iran. Whose picture would hang in every office and place of business? During the revolution many picture frames and pedestals, as well as portraits and statutes, were destroyed, but many have simply been recycled and the pale patches in the paint covered over. The prisons and military posts that used to be pointed out to visitors as symbols of the shah's oppression and as the first thing that would change have also been in full-time use since the revolution. The stories of theft and corruption by the principal clergy—except Khomeini himself—and members of their families have replaced the stories about the shah's relatives. Iranians believe now, as they did under the shah, in an improbably effective degree of government surveillance, and they are convinced of the possibility that every telephone call is monitored. Indeed, they assist in their own terrorization by exaggerating the facts of oppression. The availability of paranoid fantasy is one of the principal resources of this or any Iranian regime, for the belief in terror is cheaper than the consistent exercise of it. The government controls all the media of mass communication, but the new regime differs from the old in banning most music and entertainment as well as dissent. Both regimes have sponsored mass demonstrations and street theater of various forms, as well as calendrical rituals, though the feasts of the new regime differ from those of the old.

The Pahlavi regime used force when it was necessary to obtain assent but apparently believed that the long-term basic mechanism of assent would be the achievement of improved conditions of life, that modernization was its own best argument. The mullahs also use force, but they are striving to obtain assent at a more emotional and ideological level, and they expect to accomplish this even when conditions are notably deteriorating. The shah's vision required a gradual improvement in the economic system through greater efficiency and the optimum use of human and natural resources: his rhetoric was based on self-interest, not on sacrifice, but the Islamic republic uses a rhetoric of sacrifice. The Orwellian vision of totalitarianism suggests that tyranny will be easier to maintain under the pressure of constant scarcity, drabness, and military mobilization. Any attempt to guess the course that would have been taken by Iran without the war with Iraq is speculative, but there is some reason to argue that military adventurism was inevitable. Certainly affirmation of the need for unity and solidarity in the face of a common enemy is important in Iran today. More striking still is the promotion of inefficiency in other ways. It is more important to purge the managers of an industry who have been politically or religiously compromised than to keep those in place who can make it function efficiently. Furthermore, the dismantling of the educational system and the abrupt removal of the female half of the population from the labor force represents a truly drastic change in priorities.

Most observers who worked in Iran under the Pahlavis felt that Iran was in an exceptional position. It had rich financial and natural resources and a large population, so the only limiting factor in its development was the need to improve the quality of human resources—to train more managers, technicians, engineers, and teachers and thus produce a more productive work force. Since the need was for individuals who would be well informed and committed, educational reform was given among the highest priorities by the new regime.

During the revolution, in fact, one of the factors that prevented suppression of the growing protest was the commitment of the Pahlavi dynasty to the modernization of the work force. Telephone and electrical workers, for instance, were able to disrupt the communications system, and persons who possessed a high degree of technical skill, such as scientists and health professionals, were able to criticize the government in certain ways. They believed, quite reasonably, that they were essential to the realization of the shah's dreams and could make their protests with a degree of impunity.

With its new priorities, the revolution has stripped Iranian society of certain technical capabilities. Life has become poorer and drabber, consumer goods and services scarcer and their availability unpredictable, certain tasks and goals impossible to address. Even under the best of circumstances, this would be true for a full generation as a result of emigration and interrupted schooling of the young. But the suppression of dissent and of behavior that is believed to be immoral are given the highest priority in the new order; assent to the revolution is not based on self-interest, but on ideological or religious commitment. This seemingly irrational, antimodern attitude—this search for absolutes regardless of practicality and heedless of the consequences—has been central to the Iranian revolution.

The assent elicited by economic progress, even from those who benefited the most, was always deeply ambivalent. This ambivalence was expressed as cynicism. During the last two decades of the shah's regime, outside observers commented extensively on the importance of mistrust and cynicism in Iranian culture. In the mid 1970s, my husband and I were involved in an interdisciplinary research group formed to study this phenomenon. We found, as would, I believe, be predicted, that distrust was not completely pervasive and that Iranian social life was based on affirmations and positive values as well as on calculation and self-interest. We found that cynicism was counterpoised by a strong desire for and valuation of integrity, a nostalgia for a society in which suspicion and hypocrisy would not be necessary forms of self-defense—in short, for a society in which one's worst suspicions of others would not be continually confirmed.

A variety of techniques, including interviews and analyses of literary materials and popular films in which similar themes recurred was used in the research. Popular films repeatedly presented images of persons who were corrupt or had lost honor in some way but suddenly found the determination to risk all for some overwhelming commitment that restored them to strength and integrity. In one film, for instance, a widower who has neglected his children has become an alcoholic and is living in degradation. On hearing, however, that his daughter has been seduced, he suddenly pulls himself together, stands up straight, and heads north through the city, bent on revenge—walking down the center of a highway against the traffic. Such characters were depicted as probably doomed to martyrdom, for the films and stories that asserted the value of integrity also underlined its cost. The image of the father, restored to honor and imperiled by the automobiles whizzing past him as he walks with complete singleness of purpose, prefigured for me the images that were of great significance during the revolution—of marchers in demonstrations expecting to be shot down at any moment, yet moving forward unarmed, or of householders going up onto their rooftops to chant after the curfew, wearing the ritual garments of martyrdom.

There was a great deal of comment during the revolution about the concept of martyrdom in Shiism. The reservoir of willingness to

Mary Catherine Bateson

risk all for some ideal and the celebration and glorification of those who did so were much discussed. What was generally not conveyed was the nostalgia for such a commitment and the concomitant belief that one had compromised one's deepest values through a mixture of fear and self-interest. Great amazement has been expressed that United States officials and reporters were unaware of the level of hostility toward the shah—and, by implication, toward this country. This hostility was, in fact, the other side of a more usual grudging acceptance. One can accept benefits from a corrupter or seducer, then suddenly discover rage at being corrupted or seduced. The rage against America during the revolution was largely based on self-hate.

The revolution offered people the opportunity to shift gears, to realize the possibility of risking all for a commitment that they had always felt in themselves but had not expressed. Khomeini himself had a record of absolute unwillingness to compromise, and he sustained that record through the revolution by refusing to meet with representatives of political groups that were prepared to negotiate or work for compromise solutions. The popular preaching that fanned revolutionary fervor drew on Shiite images of martyrdom, but it also berated people for benefiting from the shah's regime. One of the most common images, for instance, was that of the man who lives off the whoring of his womenfolk, with Iran serving as the image of a mother turned into a whore. One of the frequent themes of discussion was an emphasis on the extent to which supporting the revolution was *against* one's self-interest. Professors said, "Probably the new society will invest in social justice rather than in higher education, which will hurt me, but I believe it is right." Women said, "If I believe that Iran will be truly free and social justice will be achieved, I am ready to give up my rights, and I would even accept it if my husband took another wife."

Persons who felt divided between their attempts at survival or advancement in Iranian society and their yearning for integrity found the decision to become supporters of the revolution immensely satisfying. The absolute lack of compromise that had been so appealing in Khomeini before the revolution became the basis for the absolutism and totalitarianism that have developed since. The enforcement of ideological uniformity is not a means but an end, and there is no room in Iran today for pluralism or diversity of views. The Islamic republic is not concerned with better or worse, but with good or evil.

This interest in absolutism resonates with certain characteristics of Islam, especially of Shiite Islam, but it is not specifically or exclusively religious. It turns up in a range of settings not associated with institutionalized religion, for example, in relation to masculine honor and romantic love. Nevertheless, Islam specifies behavior throughout a wide range of areas. The canonic law of Islam establishes norms of personal hygiene, food preparation, family life, public order, commerce, and government, as well as for religious belief and ritual. On many of these matters Islamic legislation is highly specific and is based on well-established texts, so there is little room for disagreement. On others, while some writers have attempted to go back to the spirit of Koranic texts to argue for alternative approaches, the position of the clerical majority is quite clear. The drive for internal consistency that seemed to lie behind much of the revolutionary fervor and the drive for external conformity to make Iran a fully Islamic society are complementary. The only kind of pluralism Islam does provide for is the possibility of second-class citizenship for members of specified religious minorities.

The revolution in Iran was to a considerable degree made possible by a desire for the absolutes, a desire that now supports totalitarianism. But this same cultural tendency, and tradition, may in time prove to be a source of weakness in the Islamic republic. Individuals, accustomed to living in what they experience as a state of compromise, can occasionally shift to a state of total commitment. But ordinary life and belief are in fact full of compromise and inconsistency, and Iranians are pessimistic about the maintainance of integrity, either by leaders or by themselves. The Iraqi war may have helped Iranians postpone the realization that the new regime, like any other, is a compromise. But it is not clear whether this insistence on conviction and dedication will ultimately prove to be a source of stability or of instability.

Mary Catherine Bateson

Authority and Tolerance: The Dutch Experience

JAN VAN DORSTEN

*"The people say and print what they please,
and call it liberty."*
—John Ray
*Travels through the Low Countries,
Germany, Italy, France*

How does one describe the Netherlands? Although Holland no longer rules the waves, Rotterdam has the world's largest commercial harbor. The warehouses of Amsterdam merchants no longer determine the world market, yet today the Dutch are the largest foreign investors in the United States. Perhaps Benjamin Franklin's remark, "Some writer, I forget who, says that Holland is no longer a nation, but a great shop," applies to any period in Dutch history. Except that he was wrong about it being a "nation."

In the seventeenth century, when the Dutch were still anxious to assert their power and their presence—in Europe, South Africa, the Far East, New Amsterdam, wherever—there was no such thing as a Dutch nation. There is now, but paradoxically, the Dutch today appear to operate more anonymously, merging with the rest of the world in international corporations and enterprises. Of course they still grow tulips, they still produce cheese and beer, and they are still interested in energy and technology. But Royal Shell has replaced the windmills, and the colorful flags of the Dutch republic have given way to more discreet forms of self-advertisement. The Dutch presence is more masked, more hidden: the winning keel of the yacht *Austrailia II,* for example, which won the America's Cup in September, was developed in the Netherlands. In all fairness, I should add that the American yacht that lost was designed entirely by a Dutch marine engineer, now a U.S. citizen. We now have a constitution—and a queen—a central administration, a nationwide bureaucracy, and all the standardizing influences of a modern state. But in order to make the regional programs intelligible to the country, Dutch television is obliged to offer subtitles in standard Dutch when it presents speakers from certain provinces that lie not so very far from Amsterdam. And regardless of the leveling influence of paper culture and mass media, local government in the North continues to be conducted in a different language, Frisian—as it always has been.

In spite of increased centralization in recent times, it is still important to keep in mind that the Netherlands that emerged as some sort of country in the sixteenth century remained for centuries a rather loose confederacy of independent provinces and towns, held together only by the need to cooperate in times of common peril. The United Provinces had only become united—more or less—because of their shared dissatisfaction with the centralized authority of their lawful sovereign, the count of Holland, who also happened to be the king of Spain. And it is revealing that at the very time when the rest of Europe was moving toward absolutism and the divine right of the monarchy, the Dutch established a republic. Prominent in it were the so-called stadholders—more than one at first—but they were neither kings nor even presidents and generally had rather less power than, say, a president of the United States.

It is impossible to explain the mechanics of such a country in modern terms. The republic had not been planned. It lacked a coherent theory of government. Very few Dutchmen looked upon the seven provinces as their fatherland. "Home" was one's region; what mattered was peaceful agriculture and freedom of commerce. The principal business of a magistrate was simply to look after the waterworks and waterways and to ensure that the roads were safe. His concern always was, and is, small in scope. In such a small territory, without any desire to build an empire, larger issues, such as a war office, that require a powerful central government generally seemed less important.

The virtues of such an unsystematic technique of governance have often failed to impress others. "The constitution of government," John Adams complained in 1782, at the time of the fourth Dutch war with England, "is so complicated and whimsical a thing, and the temper and character of the nation so peculiar, that this is considered everywhere as the most difficult Embassy in Europe." Adams knew what he was talking about. For more than a year he attempted to obtain recognition of the American republic by the Dutch republic, only to discover that in order to achieve any decision at all he had to make endless personal calls on a large number of individual politicians, plus all the representatives in The Hague of the major Dutch cities. Dutch towns, he wrote to the U.S. Congress, were like independent republics, and their mayors like kings. In the end he prevailed. The Dutch republic was the second power in Europe to acknowledge the United States as an independent nation, after which Dutch bankers were quick to provide the money to cover the considerable U.S. foreign debt after the American revolution.

That is how the system worked—perhaps how it has always worked. As Dutch politicians discover every day, the Kingdom of the Netherlands, like its republican predecessor, is nervous about authoritarian rule by powerful individuals. Decisions continue to be made circumspectly. They require time and political tact rather than power, bribes, or brains. We call that democracy. In present-day Holland, political parties abound. So do churches, sects, and religions. Radio and television are run by a variety of cultural, denominational, or political organizations. Coalition governments—what else?—are established and dissolved, though not as frequently as in Italy. How can any outsider ever be expected to assess the value of what a Dutch government spokesman says on behalf of "the nation"—whether on nuclear disarmament, apartheid in South Africa, or any subject discussed in the United Nations—if he does not know the conventions of Dutch policymaking? And why should he? What he remembers from the newsreels is, perhaps, riotous scenes from our capital, Amsterdam, with its rich tradition of controlled anarchy. But if he is a historian as well, he will also recall an equally rich tradition of tolerance.

Because of lingering Puritan influences, we have no Sunday papers, and all shops must be closed on the Sabbath. But in our more emphatically Puritan seventeenth century, foreign Protestants observed with horror that we conducted our business all seven days of the week. The majority of the founders of the Dutch republic insisted that there be room for such freedom and resisted the idea that an old tyranny should be replaced by a new one, even that of the Reformed church. And so, whatever the authoritative ideology of the Calvinist republic was, there was always room for dissenters, Jews, foreigners, freethinkers, even Roman Catholics. In retrospect, we appear to have been a tolerant country—and in comparison with some surrounding countries, we certainly were. But tolerance was practiced rather than preached. Where it existed, it existed not because of nationwide laws on tolerance, but simply because it was allowed to exist, somewhere, somehow.

The Dutch have always tolerated the presence of other persons and views as long as they did not become nuisances. The future American Pilgrim fathers could live in Leiden quite peacefully as long as they behaved well and kept to themselves. The story in American schoolbooks that they left Holland because their children were beginning to look like Dutch children is entirely unfounded and is not quite fair. The Dutch never insisted that strangers should become Dutchified. Recent attempts at enforced assimilation have been unconvincing, and as a rule we prefer to leave, say, the Moluccan communities from Amboina alone as long as they do not hijack the trains. On the other hand, you might be allowed less freedom if you were the local schoolmaster in a small country town. But that, I am sure, is universal.

Tolerance is always an ambiguous thing. Modern studies have emphasized either the powerful intellectual tradition of tolerance in Dutch humanist thought, or, quite the opposite, a purely practical habit of noninterference with others in the interest of trade. What the merchants and the intellectuals had in common was that they could think beyond the narrow boundaries of the province or state. Most merchants certainly wanted peace, or at least an opportunity to trade peacefully, even in times of war. Dutch bankers supplied money to all parties, and trade relations with the enemy have always been common practice.

But as a rule, the merchants favored tranquillity, as did the idealists of the age—the countless Christian humanist spokesmen for religious tolerance, from Erasmus onward; and the followers of the so-called radical reformation—Anabaptists, Mennonites, and spiritualists. "Should every man believe what he wishes?" is a famous question put to the late-sixteenth century irenic writer Coornhert. His answer was: "Yes. Otherwise he must believe what somebody else wishes him to believe." And in spite of much intolerant thinking throughout the centuries, this remark, with its insistence on the freedom of the individual to select his own road to salvation or damnation without the interference of "authorities," strikes me as fairly typical. It leaves room for a wide variety of attitudes—from mere indifference to the most intensely individual forms of religious faith, which can include both spiritualism and Calvinism—and it creates space for those who, rightly or wrongly, believe differently.

Characteristically, the first university in the Dutch Republic, the University of Leiden, though Protestant, required no religious oath from its students or staff, admitted numerous dissenters—and some Roman Catholics—and even used its legal privileges to protect its members against outside authorities. Characteristically, too, that same university prepared the way for the first acquittal in a witchcraft trial. In the republic the last witch was burned in 1595, more than a century earlier than in the surrounding countries.

Against this image of a tolerant, liberal-minded country, numerous examples of the most intolerant activities could, of course, be cited, beginning with the Synod of Dort in 1619, the execution of the venerable statesman Oldenbarneveldt, and the prosecution of other eminent men, such as the famous lawyer Hugo Grotius. But here again, no adequate picture of the country as a whole can be given because the enforcement of state regulations varied from province to province, and indeed, from town to town. In things that really mattered, the decrees of a central authority were always modified by local practice.

Jan van Dorsten

More freedom could be found in the Netherlands than in most other countries, if you knew where to look for it. This point is central to any discussion of tolerance in the Netherlands. It can perhaps best be illustrated by focusing on the best-known example of Dutch tolerance, the freedom of the printing press.

There was never complete freedom, but because it was fairly easy to find a suitable printer in a suitable city at almost any given time, countless books in various languages appeared that could not be printed, and never could have been printed, in neighboring countries. Some books and pamphlets *were* forbidden—perhaps a thousand during the entire period of the Dutch republic—but the ban tended to concern theological works and scurrilous books rather than political treatises. There was no preventive censorship, however, and only a relatively small number of books were banned by a national authority. As a result, a book that was forbidden in, say, The Hague, could be printed and sold over the counter anywhere else. In some instances pressure was exerted by church authorities to take legal action against a "dangerous" work, but instances in which the civil magistrates were unwilling to cooperate until the book had gone through so many editions that a ban became superfluous could also be cited. Only in the eighteenth century did the central administration feel sufficiently confident to float the idea that nationwide, prepublication censorship might be desirable. It provoked such an uproar that the idea was abandoned for good—only to be revived in times of foreign occupation, such as the five years of World War II, when some 1,000 clandestine books and pamphlets, not counting newspapers and broadsides, were published. As a result, almost every book of importance, and many more that were unimportant, could be published in the Low Countries—if you knew how to evade local restrictions.

Some famous controversial works thus appeared in the Dutch Republic. To name only a few authors: Comenius, Descartes, Hobbes, Locke, Pierre Bayle, Voltaire, and Rousseau. The men who published and sold their works were not backroom revolutionaries or romantic idealists. They were merchants who produced good books on good paper for a good market. The Dutch book trade in republican times owed its success to the simple fact that it produced outstanding quality at low prices and that it could rely on well-established trade routes—to England, for example, where censorship, trade union restrictions (by the Stationers' Company), and high wages made the import of foreign books, legal or illegal, a profitable business. An Amsterdam printer claimed in 1680 that he had smuggled a million English Bibles—the forbidden Geneva Bible—into Britain. Statistics based on English customs declarations between 1700 and 1780 indicate that 61 percent of all foreign books came from or by way of the Netherlands, and only 20 percent from France. This of course does not include the smugglers' trade, for which, understandably, we have no figures. It is clear that the Dutch book trade was a large, international, nonideological enterprise. It offered for sale anything (or almost anything) for which a demand existed or could be expected to exist, and in most cases the authorities were wise enough to let the booksellers go ahead without too much interference.

In 1941, when the Germans occupied the Netherlands, the great historian Huizinga published one of his finest works—based, incidentally, on a series of lectures he had given in Germany before the war. Not surprisingly, it was devoted to the virtues of the Dutch republic in earlier days. But this was not his last book. In the darkest months of 1943, he wrote *Geschenden Wereld* ("Violated World"), on the question of how civilization might be restored in the future. It was published immediately after liberation day—a day Huizinga did not live to see—and it argued that the greatest common evil is a state that "pretends to be the measure of all things while at the same time proclaiming its own immoral character." Voltaire, who knew Holland and had had mixed experiences with the Dutch, remarked accurately, in his own characteristic way, that "in Rome one is a slave, in London a citizen, but a Dutchman is proud to live without master."

In spite of our quondam colonies, the Dutch experience is not one of empire building, but of coexistence. It is generally unheroic and, thank God, relatively peaceful. It is also eminently practical. Perhaps this is why Holland has given the world some great painters, businessmen, scholars, and scientists, and of course sailors, but no great imaginative philosophers. When the Dutch printers enjoyed their remarkable freedom, Holland did not have a John Milton or a Roger Williams to define the principles of the liberty of the printing press. But the presses were busy.

Jan van Dorsten

MEDIA

"I am old and tired, so I am stepping aside.
From now on, television will rule your lives."

The Rise of the Computer State

David Burnham

"The true danger is, when liberty is nibbled away,
for expedience and by parts."
— Edmund Burke

George Orwell's *Nineteen Eighty-Four* is directly relevant to the world today. The computer, as marshaled by the great public and private bureaucracies of the United States and other industrial countries, is contributing to Orwell's significance as a prophet.

Of first examination, the looming presence of Big Brother and the overwhelmingly shoddy physical environment inhabited by Winston Smith seem quite distant from our well-fed, superficially glossy age, with its pluralistic babble of political voices. But to take this view is to overlook some extraordinary parallels between the two worlds.

In Orwell's 1984, society was divided into two separate camps. Winston Smith, the hero of the book, is a lower-level functionary working for the government, the nation's only bureaucracy. With the video screen in his room, the microphones, the constant surveillance, the ever present Thought Police, Smith is totally plugged into the social system of his world. Every movement, every thought, every flicker of an eye is recorded and noted by the party.

The ultimate goal of the party is to stamp out individual awareness, to eliminate individual thoughts, and to murder all forms of individual consciousness, including even the awareness of sexual pleasure.

The contrast between Winston Smith and the proletarians, the other social class in Orwell's Oceania, is striking. Although the author does not tell us very much about the proletarians, they are a looming, yet almost totally impotent mass. Orwell makes it clear that the proles have no schools, no skills, no doctors, no transportation, no security. They have no political voice. In a real sense, the proles do not exist.

This ugly, starkly stratified world of Orwell's last work has very little to do with life in the United States. Or does it?

Do not scores of government and business organizations use computers to pry into the lives of millions of Americans in much the same way that Big Brother's lieutenants followed Winston Smith? Do not the hundreds of thousands of unemployed and unemployable men and women living in the rubble of the south Bronx, or around the auto works of Detroit and the steel mills of Pittsburgh, share many of the experiences of Orwell's proletarians? And is not the computer gradually increasing the barriers that separate the two camps?

Consider the life of a successful banker in Columbus, Ohio. With his home, office, and mobile telephones all plugged into advanced computers, a complete computerized record is kept of every number he dials, how long he spends on each call, how many times he misdials, and the location from which he placed each call.

Because the executive's home is equipped with two-way interactive television, the computers of the Warner Amex Company collect information about when he enters and leaves the house, those instances when he chooses to watch one of the pornographic movies offered by the system, how he voted on local and national polls, and perhaps even what books he ordered from the library.

During the four or five hours that the banker and his family watch television each day, they are subjected to a repetitive and extraordinarily powerful barrage of advertising, cunningly designed to prevent critical thought and to block individual judgment. If he should live in certain sections of the country, his family's individual purchases at the supermarket are automatically matched with the advertisements they have seen on television. The marketing experts can thus keep track of their human guinea pigs in order to improve the effectiveness of their incessant advertisements.

When he takes a business or vacation trip, computers read his credit cards and automatically record every hotel in which he sleeps, every meal that he eats, every rented car that he drives, and how far he travels.

But that is just the beginning of the surveillance system that keeps track of the movements and thoughts of millions of Americans. An increasing number of companies, for example, have taken to maintaining a computerized record of exactly whom every person in the firm reaches on the interoffice telephone system, how many minutes each person spends at the word processor, which motor pools he or she joins, in what corporate social activities he or she takes part, and how much he or she has contributed to the United Way.

So much for the private sector. The Internal Revenue Service is now testing a surveillance system for which it will use computerized lists of individual households and their estimated incomes, developed by commercial mail-order companies, to identify those who pay no taxes or too little in taxes. At the same time, the Federal Bureau of Investigation seeks to push through a project aimed at enlarging its electronic surveillance system by automatically keeping track of the "known associates" of drug traffickers and a number of other categories of people who are not wanted for specific crimes. Meanwhile, the Selective Service System is using the computerized information that every American must supply to the IRS to send warnings to young men who fail to register for the draft.

These federal surveillance systems are in place today. Within a decade or so, if the Federal Reserve Board has its way, checks and cash will gradually disappear from the U.S. economy. Checks are too expensive to handle and cash, the Board contends, is easily subject to loss or theft. As a substitute, the Board is actively encouraging the development of an electronic funds-transfer system in which everything would be paid for by electronic messages transmitted around the United States by means of the Board's computerized telecommunications network. As a senior official of the Board said several years ago, the only people who want cash are crooked politicians and illegal drug dealers. Already, the successful executive does a good deal of his personal banking in front of an electronic teller that automatically records the exact minute and amount of each withdrawal or deposit.

I call the rapidly growing collections of mechanized notations about everyday events in the lives of a significant portion of the American people transactional information. Looked at separately, these millions of bits and pieces of transactional information seem benign. The automatic collection, storage, and retrieval of so much information about so many people at so little cost, however, is gradually bringing about a fundamental change in American society. With very little thought, the power of all levels of law enforcement is being enhanced while one of the most important principles of the Constitution is being undermined.

One reason for this is that vast quantities of transactional information are now easily available to the police. Investigators are thus able to move back in time as they have never been able to do before, in order to trace the past activities of millions of citizens. A second reason for the growth of the power of law enforcement is that the existence of computerized transactional information allows the police to develop profiles of expected behavior and initiate an investigation of anyone who does not adhere to the norm.

The ability to do mass sweeps of large segments of society appears to be subtly undermining the Fourth Amendment protection of the right to privacy. The intent of the Constitution is that before a police agency can break into a suspect's home or listen in on his telephone conversations, it must obtain a search warrant from an independent party, a judge. But computerized transactional information is making this rule of law obsolete.

The surveillance of the computerized agencies of the United States is obviously far less heavy-handed than that described by Orwell in *Nineteen Eighty-Four*. But it can be overwhelming. I refer skeptics to the series of reports of the activities of the CIA, the NSA, and the IRS, published in 1975 by the Senate Select Committee to Study Governmental Operation with Respect to Intelligence Activities.

But where are Orwell's proletarians?

Go look at the people living in the south Bronx, in the South Side of Chicago, or in the distant reaches of Los Angeles or Houston, or, more and more, in Detroit.

Consider the quality of their schools or the number of police officers patrolling their neighborhoods. Examine the latest political polling techniques, using high-speed computers to identify the census tracts that the politicians systematically ignore because they have the lowest percentage of active voters. Ask yourselves about the unemployment rates in these areas, and try to determine what efforts, if any, have been made to retrain these social castaways.

Because a relatively small proportion of blacks have moved into top management and professional positions, it is sometimes forgotten that as a group they are among the country's poorest, least educated, and worst fed. In 1982, unemployment of black men sixteen years old and older was twice as high as that of white men. The rate of imprisonment of blacks is four or five times as high as that of whites. Statistical studies suggest that during the last two decades there has been a greater increase in all kinds of chronic illnesses, ranging from cancer to diabetes, among black men than among white.

Today, only a relatively small proportion of the American people have been found unfit to take part in our society. But many social scientists are deeply concerned that within the next decade computers and computerized robots are going to have a serious effect on the employment of a far larger segment of the population.

Almost all agree that hundreds of thousands of workers will have to be retrained for new jobs as their old ones become obsolete. Many others believe that millions of workers will become permanently unemployed. Massive technological unemployment, they fear, will not be limited to a relatively small number of blue-collar workers in the automobile and printing industries. Advanced computer systems are already replacing engineers, architects, office workers, and medical specialists such as physicians who read X rays. The ranks of these unemployed and unemployable, these men and women without money to own telephones or to be hooked into two-way interactive cable television, these citizens whose political views are systematically ignored, could develop into the American version of Orwell's proletarians.

We must also consider another central concern of Orwell's: the vital importance of clear language and straightforward laws.

Representative democracy cannot survive without clear language and articulate law. And George Orwell, who hated dictatorial governments of both the left and right, supported representative democracies wherever they flourished. If we cannot understand the law, if the government is able to interpret the law in any way it wishes, if citizens cannot easily grasp what is required by law, and if laws are not drafted to deal with important subjects, then representative democracy is in trouble.

If a society has irrelevant laws, if a society chooses to ignore laws written by its legislators, if a society tolerates the existence of important institutions that operate outside the law, and if the legislature denies petitioners the protection of law, then that representative democracy is seriously endangered.

In this light, consider the following:

● *Irrelevant law.* The 1938 wiretap statute makes it a crime for anyone to record a conversation secretly. There is an exception for policemen, who can eavesdrop if they obtain judicial warrants. Because of the way the law is written, however, it is not a crime for anyone to intercept and record computerized information. Recent changes in technology have thus created a situation in which a huge part of our communications transmissions have no legal protection.

● *Ignored law.* The Federal Privacy Act states that government agencies may not use information collected for one purpose for another purpose unless they inform those whom the information concerns. This is an essential provision if the Fourth Amendment is to survive. First under President Carter and Secretary of Health, Education, and Welfare Joseph Califano and later under President Reagan, this provision has been ignored. Using a questionable legal loophole, the government has moved into full-scale and widespread computer matching, regularly using information in ways that were not anticipated when it was first volunteered.

● *Outlaw.* If an institution is allowed to exist outside the law, beyond the normal checks and balances of representative government, that institution is by definition an outlaw. Such is the National Security Agency, the secret, massive, computerized Department of Defense operation that conducts electronic surveillance all over the world. The NSA, unlike the Department of Agriculture or the Central Intelligence Agency, has no legal charter. It operates under a still secret executive order signed by President Truman. The federal courts, in case after case, have refused to consider whether its surveillance violates the Constitution. In the only known case in which a federal district judge ruled that the surveillance of a Detroit lawyer by the NSA violated the Fourth Amendment, a court of appeals panel reversed the decision and held that the agency could properly intercept the lawyer's overseas messages and give them to the FBI and other law-enforcement agencies, even though there was no evidence that he was a spy or a criminal. The Congress provides no informed oversight of the NSA. Although the Constitution holds that from time to time there will be a public accounting of federal spending, the budget of the NSA is one of the deepest secrets of the government.

● *No law.* When a serious legal problem is discovered, the legislature of a representative democracy seeks to resolve it. A few years ago, a lawyer in Los Angeles discovered a secret computerized list in which landlords stored gossip and other information about tenants. This particular lawyer was denied an apartment for several months because the landlords' computer had recorded incorrect information about him. After discovering the problem the lawyer asked the state legislature to pass a law requiring landlords to inform a rejected tenant about this system, to allow the tenant to see the negative information, and to permit the tenant to offer his version of the reported event. Because of the great power of the real estate lobby, this reasonable and responsible proposal was rejected.

Does Orwell's *Nineteen Eighty-Four* present important warnings to the American people and the legal profession today? As I see it a thoughtful reading of his book and a careful examination of our society can produce only one conclusion: It does.

Computer Babble and Big Brother

ELIOT D. CHAPPLE

*"History tells us of innumerable retrogressions, of
decadences and degenerations. But nothing tells us that
there is no possibility of much more basic
retrogressions than any so far known, including the
most radical of all: the total disappearance of man
and his silent return to the animal scale."*
—Ortega y Gasset

If thoughts are those bits and pieces of an unending stream of
information of which we become aware, consciously and uncon-
sciously, at varying intervals of day and night—recognizing that the
process rarely ceases during our slumber—then clearly their sources,
outside of actual events, are derived largely from the media. All kinds
of cultural habits and predilections are built out of their ingredients—
a trip to Kenya or Las Vegas, a tantalizing recipe for a chocolate
soufflé, the latest style in clothing or the newest secrets about the
well known, together with a half-realized journalistic record of local,
national, and international goings-on. To say that this flood is media-
controlled overvalues the process and obscures the multitude of
contributing inputs of our information age, whose products, moment
by moment, form a hardly digestible lump of confused ingredients.
These pass us by, quickly enough, with new constituents crowding
in to take their place.

Of course, some elements tend to form small aggregates around
which others colalesce, like a branch lying across a stream and creating
a temporary dam behind which other driftwood can become lodged.
But with a change in the velocity of the flood, or the happenstance
of other shifts in the location of such entrapments, the whole barrier
can easily be swept away. Does the Word, as print or photograph or
electronically produced speech or film or TV image, only a single
particle in that flood, shape the way people act and interact? Such an
interpretation assumes that thoughts are the harbingers of action
rather than their consequence. Only if this order held true would it
be a simple matter for those skilled in the arts of the media to
substitute one set of symbols for another.

Long ago neurosurgeons and physiologists showed that the symbols, or the complex configurations into which they become linked, were manifested after the action had been triggered by external stimuli, even if only by a few milliseconds. When it comes to actions or reactions that we call emotional or behavioral—correctly, changes in the autonomic-endocrine-somatic nervous system—words are mere signs, small parts of the interactional contexts of the situations in which they are used. Their meanings fluctuate with the moment-by-moment changes in what goes on. What people say is an uncertain indicator of what they will do. St. Jerome said to his fellow churchmen, "Do not let your deeds belie your words, lest ... someone may say to himself, 'Why do you not practice what you preach.' "

Since the problems presented by the ambiguities of meaning, which we too often dismiss as mere semantics, are fundamental to human animals, whatever their degree of cultural and technological sophistication, we should be hard put to take words as anything more than the mouthings of the "bander log." Some of you may be old enough to remember how millions of young people in Great Britain subscribed to the Oxford Oath during the late 1930s, swearing never to take part in war, just or unjust. A few short years later, Dunkirk drove home the unrealistic nature of such thoughts.

The number of identifiable meanings, selected solely from the literature, of any common terms can easily approach the hundreds, and the shapings that each person gives to any one multiply the meanings further. Symbols carry with them a curious and complicated baggage of definitions. Their meanings shift, even within days or hours. The contexts in which they appear, moreover, vary from one situation to another, changing their composition as time goes on, becoming highly differentiated between one person or one subculture and another.

What has confused the issue and made media control a kind of bogey man of culture is the belief that symbols have specific, uniform meanings and that identification and use of specific symbols function with the relentless uniformity that we attribute to the conditioning process—the tethered animal salivating at the sound of the bell. That Pavlov also discovered that large groups of animals could not be so conditioned is ignored. The Skinnerian belief in behavioral conformity as an inevitable consequence of repetition is even more misguided. But it is on such uncertain grounds that the threat of media control is adumbrated.

What is decisive is neither input imagery nor symbolism; the central nervous system is little affected by that flood of inputs I referred to at the beginning. Direct experience only enhances the readiness to react to or to ignore whatever general significance the

Eliot D. Chapple

images and symbols contribute. The old seaman is not overcome by the onset of a gale, or the warrior by another battle. History—and our own experience—tells us that human beings are remarkably capable of shaking off events that engage them only peripherally and of differentiating between direct effects that literary people and psychologists claim should be completely destructive.

In general, experience and its repetition enable people to learn to accommodate themselves to crisis, not because they become blunted in their capacity for adaptation, but rather because their survival skills are increased. By and large, we are able to take the major crises of life with reasonable resilience and with a continuing capacity to maintain or to reestablish our emotional-interactional balance. We live today, in today's situations. Past and future are, fortunately for us, always going to be taken into account on a secondary level.

Those concerned with Big Brother, mass media and communication, and computer technology, never face up to the complexities that their programmatic apocalypses require. Assume that all we need to do is to manage the interactions of 100 individuals and, without considering the dynamics of formation of triples or quadruples, we limit ourselves to pair relationships. Even so restricted, the number of pairs of interacting individuals we need to control is 4,950, the formula being $N(N-1)/2$; if we extend our population to 1,000, we end up with 49,500 pairs. And such tiny population groups are of minimal relevance to the establishment of overall controls for world dictatorship. We need not do more than make brief historical reference to all those kings and priests and their determined efforts to condition their subjects from cradle to grave, with remarkable lack of success—except for very short periods when the interactional environment was dichotomized and unstable, as it is in Iran today.

One of the serious pathologies of this age of communication can best be described as elephantiasis of the data base. Without inquiring why it is that any conglomeration of undigested and incompatible data needs elaborate computer programs for its manipulation, database management is being strongly merchandised by the think tanks of universities and the software companies. We are apparently expected to believe that any aggregate of miscellaneous data, any item that can be assigned a position in a sequence—even a number—as on a Hollerith card, will lead us to the management of individuals against their will and knowledge, by clustering all kinds of information about them into blocks. This is the latest threat that has been dreamed up by communication alarmists. Whether the information comes from charge accounts, records of telephone calls, or social contacts in bowling, contriving such blocks can be used to give the same label to very different people and then, with Big Brother on the job, using these blocks to spot dissidents, pay off the presumed faithful, and so on.

Since when have despots needed sufficient information to build a case rather than to clap a suspect into prison, extract a confession by torture, or simply fabricate the most convenient story? Alternatively, with whatever degree of caution appears necessary, let them disappear, rot in jail, or be assassinated? If the dictator is concerned with his place in history, there will always be scribes available to rewrite it to glorify his rule.

Even today, in what we read about the customary practices of governments and rebels alike, such individual targets may be too few to satisfy a dictator. How simple and gratifying it must be to destroy a whole village or a city and all its inhabitants, or to exterminate every member of a tribe, a cult, or an alien minority! Why bother with electronic niceties of eavesdropping? What becomes essential is to eliminate all those who deviate at any moment, today or in the past, or might appear to contemplate such a possibility, using whatever criteria of "proper" action or speech the dictator may define as permissible.

In a kind of mixed-up census-demographic way, the alleged capacity of computer block analysis to identify heretics reminds me of J. L. Moreno, the founder of sociometry. By asking girls who their friends were at the reform school in which they were incarcerated, he predicted quite well which ones would try to escape. On the basis of actual observable exchange of goods and services, this idea was used by rural sociologists in farm communities during the 1930s. And long before, at the beginning of time, in Sumer say, such criteria told the wise who had influence.

Perhaps the proponents of block modeling would find their invention less threatening if they realized that dictators hardly require such computer block analysis to operate successfully. In addition, their efforts to encourage the blockees to so regulate their contacts and management of paper as to avoid becoming targets illustrate the remarkable value of computers in this world of information and communication. Rather than talk about CAD (computer-aided design) and CAM (computer-aided manufacturing), we can now have CAN (computer-aided numerology) and CANT (computer-aided numerological tautology).

The Mass Media and Their Audiences: Interconnections and Influences

ALBERT E. GOLLIN

"The fault, dear Brutus, is not in our stars,
But in ourselves, that we are underlings."
—Shakespeare
Julius Caesar

It might be useful to recall that the concern with the effects of the mass media is rooted in the apparent success of propaganda efforts made during World War I and subsequently by Nazi and Soviet regimes to mobilize, coerce, or control their own citizens. More recently, the agenda of concerns has broadened, without wholly losing the edge of anxiety that characterized discussions in that earlier era. Here are just a few of the questions that have been raised:

● Has the graphic treatment of sex and violence by the media contributed to a decline of morality and trivialized or vulgarized significant aspects of human experience?

● Has the aggressive handling and criticism of political and economic elites by the media eroded their mandates for leadership and led to a general decline in the perceived legitimacy of social institutions?

● Are the media persistently exploited for political and commercial purposes, selling us candidates and products we would not otherwise buy?

● Have the media created a popular culture that has steadily cheapened public taste: sitcoms and soap operas instead of Shakespeare and Verdi, Harlequin romances instead of Hemingway?

● Did the news media drive Nixon from office, and did they cost us victory in Vietnam?

The list goes on and on. It might be noted in this regard that the criticisms and questions raised are neither internally consistent nor devoid of special interests.

Evidence from research into mass communication is a basis for commenting on several assumptions commonly made by critics of the media and by others who believe in the power of the media to affect our thoughts and actions and to shape our society in various ways, both good and bad.

The first of these assumptions is the equating of the content of the media with its effects. In this view, what people see, read, or hear, especially when they are exposed to it repeatedly, actually has the effects hoped for or feared. Based on this simplified stimulus-response conception, for example, are the following convictions:

- Violence in children's television programs leads to violence on the playground.

- Sexually permissive norms highlighted in films, on television, or in books and magazines are echoed in the behavior of those exposed to them.

- Sympathetic portrayal of minorities generates compassion and tolerance.

- Media-based campaigns to persuade people to reduce their consumption of energy or to lead healthier lives will yield socially desirable results.

Linked with the equating of content with effects is the assumption that the intent of the communicator is faithfully captured in the responses of those exposed to the message. Thus, according to this view, M*A*S*H not only entertains, it also conveys the antiwar intent of its producers. Or Archie Bunker's bigotry, rather than giving sanction to prejudiced attitudes, is perceived as misguided, out of date, and morally reprehensible.

The evidence from communications research, while admitted to be uneven and less than conclusive, nevertheless reveals relations between the content or intent of media messages and their effects that are far more complex and variable. People bring to their encounters with the mass media a formidable array of established habits, motives, social values, and perceptual defenses that act to screen out, derail the intent, or limit the force of the content of media messages. The media certainly do affect people in obvious and subtle ways. But no simple one-to-one relationship exists.

While media audiences are massive in size, moreover—a pre-condition for mass persuasion—they are socially differentiated, self-selective, often inattentive, and, in general—to use a term once employed by Raymond Bauer—obstinate. They are elusive targets, hard to please or to convince. People actively use the media for a wide variety of shared and individual purposes and are not readily used by them. Surely, when it comes to each of us personally, we need no reminder of this fact. Why is it, then, that we believe others in the viewing or reading public to be more gullible or passive than we are?

A second assumption often held is that the mass media now operate without restraints and that their autonomy is a prime source of their power. But members of the public are not only individually resistant to the content offered them, in free societies they significantly affect content through the operation of various feedback mechanisms. In this connection, one has only to recall the decisive effects of broadcast ratings, box-office receipts, subscription and circulation revenues, and the like as market forces that constrain the predilections of media operators and producers. To these bottom-line influences that emanate from the pattern of choices made by consumers must be added a constant stream of criticism, letters and phone calls, self-criticism on the basis of professional values that include service to the public, legal restraints, and the results of marketing studies that seek to discover public tastes, preferences, and needs.

Thus, in various ways, both direct and indirect, the public acts on the mass media rather than simply being influenced by them. And with the variety of choices of content and opportunities for exposure expanding steadily, thanks to new technologies, the likelihood of mass persuasion by the media diminishes still further.

As each new item of technology emerges, it is often met by either or both of two sharply contrasting reactions. The first of these is aptly symbolized by the image of the cornucopia. The new technology is hailed for its potential benefits—enriching people's lives, removing burdens, and contributing to human progress. The contrasting perspective is symbolized by the image of the juggernaut, the machine that is irresistible or crippling or constrains human freedom.

Most technologies, the mass media included, rarely fulfill either set of extravagant hopes or fears. As they spread and become integrated into society, they change things as they extend human capacities. But so, too, do new forms of art, laws, scientific knowledge, wars, and new modes of social organization. Only with hindsight, and often with great difficulty, does it become possible to assess which of these has affected society, and human freedom, most broadly and decisively.

While at times unquestionably guilty of harmful excess and of error, the mass media are less powerful and less autonomous than their critics fear—or than their own agents sometimes like to believe. The public is far from compliant or passive and is becoming less so as media choices multiply. Finally, things are not in the saddle, riding humankind. Given the existence of diversity among the media and continuing feedback from the public, the risks of media-fostered political and cultural hegemony remain small. In any case, I suggest, such risks are inseparable from those intrinsic to the functioning of free societies, in which the mass media are now indispensable in a variety of ways.

Media Darwinism

RICHARD M. RESTAK

*"The media have brought about a fundamental schism
in our emotional lives. What we 'think' and 'feel,' how
we evaluate the world around us, the basis for our
beliefs as well as the measure of our opinions of other
people—all are increasingly dependent on our
experiences with video imagery."*
—Richard M. Restak
The Self Seekers

I am interested in some of the subtle effects of information on brain activity. I am speaking of the blunting effects of repeated vivid depictions of scenes of horror, the numbing effect of exposing large segments of the population to provocative scenes about which they can do little or nothing. The tragedies in Beirut and Grenada are a case in point. Not only were we all exposed to the agony and pain through vivid pictures, we were also introduced to relatives, who were visited in their homes and put on display in all their helplessness, frustration, and bewilderment.

The media, as a matter of policy, approach people at their weakest moments in the attempt to provide us with so-called human-interest material. The term is actually a grotesque misrepresentation. It is not human-interest material as much as it is the exemplification of a special form of evolutionary theory that I call media Darwinism: the survival of the starkest, the most intensely stimulating, the most extreme, the most revolting, and so on. This material has a dehumanizing effect on the public, while at the same time it is used to manipulate the same public into assenting to the social and political persuasions of the media.

As a case in point, selected at random, consider this sentence from the 27 October 1983 *New York Times* article "TV Reports and Debates on Crises," by John Corry: "On ABC's 'Night Line,' Lynn Sherr tried hard to get a father and mother from Morristown, New Jersey, to compare Lebanon to Vietnam.'" The parents refused. The father responded, "Peace comes very high and we know someone has to pay with their lives." I am sure that this reply didn't please the interviewer. As is usual in such interviews, these parents were shown in all their anguish, tension, anger, and frustration. I suspect that many of the parents who have been interviewed under such circumstances will regret later having agreed, under duress, to bare their souls in exercises that appeal to the emotional voyeurism in all of us. Of greatest concern to me is the effect of such efforts on our collective mental health. Repeated exposure to horror, duress, and extreme emotion stimulates one of two responses.

On the one hand are those—a group in which I would include many journalists—who literally cannot get enough of what I call the emergency emotions. Everything else eventually becomes boring. At the other extreme are those—whose response is, I think, far more common—who simply shut off, drop out, or erect some kind of barrier between themselves and what they see on television. The result of these two diametrically opposed processes is the combination of voyeurism on the part of the media and indifference on the part of the public that seems to characterize much of present-day life. Journalists speak again and again of the apathy of the public, which they attempt to overcome by resorting to ever starker and more graphic images of horror, tragedy, and so on. The public then withdraws further, and this withdrawal is interpreted as additional evidence of indifference and apathy.

Ironically, if a totalitarian regime were to design a program for controlling the population of an enemy country, it could hardly improve upon the situation that now exists in the United States:

- Media that are preoccupied with presenting vivid images of stressful and upsetting events about which the viewer can do nothing and to which he therefore can respond only passively.

- Media that are polarized in espousing leftward, liberal policies under the guise of objectivity and are doing whatever is necessary to persuade rather than inform. I refer again to the Lynn Sherr interview mentioned earlier. Ms. Sherr and many other journalists operate under the simplistic assumption that each event is merely a replay of something else. Thus, despite an incredible number of differences, Lebanon and Grenada are each seen as merely "another Vietnam." Those who do not hold this view are not likely to be interviewed; if by some chance they are interviewed, their statements are likely to be edited out before the interview is broadcast. "We simply haven't enough time to put forward every point of view. We have to be selective." This selectivity typically means putting forth the point of view favored by the journalist.

It is vitally important that the media begin to cooperate in an effort to develop new ways of keeping the public informed about happenings in a world that is growing ever more violent without, at the same time either contributing to that violence or numbing everyone into a state of psychic indifference.

Richard M. Restak

Can the Mass Media Control Our Thoughts?

Douglass Cater

"Religion is the opium of the people."
—Karl Marx
*Critique of the Hegelian
Philosophy of the Right*

"Black is white, white is black."
—George Orwell

Today we witness a mighty struggle between government and media professionals over the control of words and images. Not wishing government to dominate this struggle, we are repeatedly stimulated to preventive actions. At the same time, we ignore the fact that we have entered a new age of communication. Television is a medium that reaches the masses to a greater extent than print media ever could. We refuse to consider the implications of this revolution, or else we conduct inquiries that would beggar the capacity of Jonathan Swift to satirize.

A surgeon general's commission spent years reaching the conclusion that "there is preliminary and tentative evidence of a direct causal relationship between televised violence and subsequent agressive behavior on the part of *some* children." The learned social scientists declined to speculate on the number of children that might be involved, in an age when violence has become endemic. Television authorities, on the other hand, roundly denounced these findings— while steadfastly assuring their advertisers that the thirty-second commercial has a proven effect on public behavior.

Television has taken us a great distance toward the society in which the abbreviated and superficial transmission of words is the dominant organizing force. Profit—not ideals of truth or concerns of governance—sets the priorities of the organization of our words by commercial television.

We cling to the stereotype that the informed citizen is the ultimate sovereign in our society. Yet we do not ask the most elementary questions about how the citizen can be adequately informed about the apocalyptic issues of our time. A major network devoted four successive midnight hours to a program entitled "War Game." It simulated the dialogue of the National Security Council after a hypothetical invasion of northern Iran by Soviet troops. So enticing was the lure of network television that top officials of former administrations played the roles of president, secretaries of state and defense, and other chieftains. Their deliberations were carried to the watching millions in Saint Vitus's dance rhythm: two or three minutes of discussion round the table, another minute or two of analysis by a prominent anchorperson with consultant, ninety seconds devoted to commercial messages, and then another go-around. In such fashion we followed the dispatch of American troops to south Iran and the guarded debate over "employing the nuclear option" once it became obvious that ground troops would not suffice. This was super television, stirring all our basic passions while allowing ample time for selling products.

I would be interested in a postmortem that would show exactly how much we, the people, learned. It was show biz, pure and simple. It transformed serious leaders who have had real experience with statecraft into truncated character actors in a jerky pseudodrama. Perhaps the more pernicious influence of television as a medium of communication is what it does to those men and women called leaders who feel obliged to accommodate to its organizing principles, who rise at dawn for brief exposure on the early morning show before being hustled off the air by the impatient commercial.

Walter Cronkite now laments the growing influence of show biz on the nightly television news. Yet we must ask whether news—even if it could be restored to a pristine state—can adequately serve the function of informing the people in a free society. Long before Orwell, Walter Lippmann addressed this question. In his classic *Public Opinion*, written in 1921, Lippmann described the function of news as that of a roving spotlight intermittently illuminating the landscape. He wrote that the function of truth in society is "to bring to light the hidden facts, to put them into relation with one another, and to create a picture of reality on which man can act." This is the important question we must face: not Can the mass media control our thoughts? but Has the communication revolution brought us any closer to attaining a picture of reality on which men—and women—can act?

EPILOG

George Orwell, drawing by David Levine

Orwell as Friend
and as Prophet

T. R. FYVEL

*"But thought's the slave of life, and life time's fool;
And time, that takes survey of all the world,
Must have a stop."*
— Shakespeare
King Henry IV, Part I

Soon after World War II there came across my desk at *Tribune*, a socialist weekly of which I was editor, a reprint of John Habberton's best-selling American novel of 1881, *Helen's Babies.* I dare say that few people today remember *Helen's Babies*—or for that matter John Habberton—but in its day his light, humorous period piece about a respectable bachelor in American middle-class suburbia left in helpless charge of his two obstreperous small nephews was more widely known than the work of Emerson, Thoreau, or Whitman. As a late-nineteenth-century best-seller it swept the United States and Europe.

Orwell and I had both read the book as children and had been fascinated by the adventures of Budge and Toddy and their bachelor uncle (I had read *Helen's Babies* in Switzerland in German translation). We both recalled our childhood vision of their colorful and genteel milieu quite clearly. As a student of popular culture, Orwell always maintained a special interest in light best-sellers.

At any rate, when in 1946 I gave Habberton's book to Orwell, he returned with one of the nicest of his short essays, one entitled, after the American ditty, "Riding Down from Bangor." This short essay reveals important aspects of Orwell's outlook. Going back to his childhood reading of American popular books, he described the nineteenth-century America reflected in these books as capitalist civilization at its best. He saw it as a society which, with all its defects, was still largely innocent, transfused with a sense of buoyancy, freedom, and expansion. This was before the cultural fall that Orwell associated with the massive American industrialization which was to follow. As for John Habberton's respectable middle-class Americans of 1881, Orwell wrote that, in spite of their genteel absurdity as characters, they possessed an asset he called good morale and a basic optimism about the future that the post–World-War II society of his day had largely lost.

Was nostalgia for childhood visions of a colorful past mingled in Orwell's pessimistic view of Western society of the 1940s? Of course. We are all dreamers, and Orwell's dreams, his own early childhood memories, whether tragic or idealized, persisted to the end. From American popular books that were read by children of his Edwardian generation and that he thought more colorful than their English counterparts he had retained an idealized image of a sunlit American past. It was in part this child's image that he set against the reality he thought he saw when he sat down to write the prophecy of *Nineteen Eighty-Four.* Politically, to be sure, he was strongly pro-American. The society of the United States that had emerged victorious from World War II and in whose popular culture his eye lit upon horror comics and pulp fiction, however—this society filled him with unease.

One can picture Orwell as a prophet on the remote Scottish island of Jura in 1947, as he prepared to write *Nineteen Eighty-Four.* One can see him as a writer who looked back with nostalgia at his lost Edwardian childhood and saw virtues as well as defects, a writer who thought his English upper-middle-class world was being swept away, a writer seeking to look at this new society in a nuclear age critically and with clear eyes. *Nineteen Eighty-Four* is really a dark cartoon—the brilliant, deliberate cartoon of a totalitarian society in which an all-powerful party rules forever, and the feared image of Big Brother is projected forever, and history is falsified forever; where, in fact, all the characters engage in unchanging diabolical antics and are forever observed and frozen on the telescreen.

Orwell said that this cartoon of a book was a warning against what could conceivably happen if men relaxed their defenses against the forces of totalitarian evil. Although some took it as such, Orwell said that it was no direct attack on socialism, least of all on the democratic British Labour Party. In an essay on his friend Arthur Koestler, Orwell observed that the purpose of socialism was not to bring about Utopia but to make society just a little more egalitarian, just a little better.

This was Orwell's attitude as a dutiful British citizen. But as an international prophet who had witnessed the ravages of war, concentration camps, and atomic warfare, his vision was different. I believe that at its core lay two simple if unfashionable thoughts. The first was that various international forces which had emerged during and after the war were threatening a deterioration of human society. Second, he saw the defense against this deterioration neither in blind conservatism nor in the proclamation of false Utopias but only in our traditional democratic values: true freedom of speech and rule of law, precise and truthful political language, and above all, common decency

---

in political and social relations. Out of such ideas he fashioned *Nineteen Eighty-Four*. It is a warning that has become a parable with which to measure contemporary history.

How did this happen? Those who have not read Orwell for some time should be reminded that he was a marvellously inventive writer. In his early novel *Keep the Aspidistra Flying* he grudgingly lets his antihero Gordon Comstock be "good with words," and so was Orwell. There is an enduring quality to the very titles of his books—*The Road to Wigan Pier, Homage to Catalonia, Animal Farm, Nineteen Eighty-Four*. There are two levels to Orwell's writings. *Animal Farm* is a parable about the way all revolutions end with the substitution of one ruling class for another and that this is especially true of the Bolshevik revolution. But it is also a beautifully worded fable. One thinks not only of the inevitability of such a phrase as "some animals are more equal than others" but also of Orwell's precise, evocative descriptions of rural life, which have allowed millions of children to read *Animal Farm* as a straight fairy tale.

If one thinks of *Nineteen Eighty-Four* it seems remarkable how much of the imagery of this anti-Utopia has entered into the language— how readily people refer to Orwell's conceptions of Big Brother, of the daily Two Minutes Hate, of political facts disappearing down the memory hole, of the marvelous linguistic invention of Newspeak. Was there something prophetic in Orwell's creation of this imagery? Of course there was.

I think this prophetic touch was linked to Orwell's common-sense approach to current political thought. He hadn't much time for "the God that failed"—he had never believed in communism. In his essay on Arthur Koestler, Orwell said that the sin of left-wingers from 1933 onward was that they wanted to be antifascist without being antitotalitarian—and for him it was as simple as that. Again, when writing in 1940 about Charles Dickens, referring to the infinite complexity of human interaction and the built-in tragedy of the human situation, he said that Dickens as a nineteenth-century liberal lacked the coarseness to think as Orwell's contemporaries did—the coarse-ness that led people to believe that all social ills could be cured merely by altering the shape of society. Again, his comment seems simple common sense.

Orwell was not concerned with predicting a future he was destined not to see. Our Western consumer society of today, enjoying material plenty, may seem a far cry from the totalitarian poverty of the society depicted in Orwell's *Nineteen Eighty-Four*. Hitler and Stalin, the giant ogres who were Orwell's models for Big Brother, have curiously shrunk in size and significance, no longer terrorizing us. And yet a new look at the intuitions of *Nineteen Eighty-Four* reveals enough striking prophetic touches to explain why the book has survived so well as a political parable.

Orwell pictured a future dominated by three totalitarian super-states always at war with each other—not directly, but in changing fringe wars in the soft underbelly of the south, with Britain as America's Airstrip One. If we look at the big-power rivalries of today—at the American involvement in Vietnam and Central America, at the Soviet involvement in Afghanistan and Soviet-Chinese frontier squabbles, and at British popular protests against the stationing in Britain of American tactical nuclear missiles—the shadowy picture of the future that Orwell sketched in 1948 does not seem a bad guess.

A criticism of *Nineteen Eighty-Four* has been that Orwell's totalitarian party wielded power not for any ideology but simply for the sake of wielding power. Yet a look at the Soviet Union of today, reveals that the Marxist-Leninist ideology which originally inspired the revolution is quite exhausted. No one believes in it anymore. Even so, the totalitarian rule of the Communist party leaders goes on simply because it has to go on; KGB violence goes on because it has to go on; thought control remains because it has to remain; and history is still falsified as mechanically in Moscow as in Orwell's Oceania.

The satire of *Nineteen Eighty-Four* was directed primarily at the Soviet Union and was taken as such. References to Orwell are found in the works of leading Soviet writers, and Russian friends have expressed to me their astonishment that this English writer should have understood so accurately the Communist society he had never seen and from which they themselves escaped decades later. In *Nineteen Eighty-Four,* however, there are also ideas that should concern us in the West, such as the picture of inner-city decay, the supervision of all citizens by means of the telescreen, and the attempt to control all thought by changing current language into Newspeak. For in *Nineteen Eighty-Four* Orwell also depicts Britain and the United States as parts of a single totalitarian empire. This does not mean that he expected the two countries to become totalitarian. It can, however, be taken as a warning against certain trends he thought he saw at work in shaping our technological Western society.

It has been said that he looked on these social trends too pessimistically. The picture of utter urban squalor that he drew of London in 1984, for example, obviously reflects his impression of the still heavily bomb-damaged and war-exhausted London that he carried away with him to Jura in 1947. Well, Orwell was never a man for the clean and leafy suburbs, but even so, have we not become troubled in Britain and the United States by the multifaceted human blight that has overtaken the centers of our larger cities?

Again, take Orwell's image of every citizen potentially watched all the time by the party through the telescreen. Of course no such situation has been reached, but do we not hear of growing unease in Germany, Britain, and the United States over the fact that more and more personal details of our lives are anonymously indexed by the state in a computerized memory-storage system? Or that these details are in the possession of impersonal agencies which are consulted about our creditworthiness? In other words, that all unbeknown to us, we *are* being watched?

Newspeak may be an ingenuous caricature: But have we not become uneasy at the unceasing efforts of television advertising professionals to substitute thought-control slogans extolling products for thoughts about them? Or at the fact that these same techniques of repetitive television slogans are being applied to politics—especially in the United States?

It is the métier of a prophet to be pessimistic, and as I look back, I can think of some reasons for Orwell's pessimism. First, when Orwell began to write *Nineteen Eighty-Four,* Hitler's demonic dictatorship had only been shattered two years earlier, while Stalin's total domination over the vast Soviet Union was still in full force. To write about a final dictatorship by Big Brother did not seem so fanciful. Similarly, the atomic explosion over Hiroshima was still a fresh memory in Orwell's mind. It was not a very promising time in history.

Second, I think that while Orwell was himself a calm atheist, he was always watchful for the wider social effects of the decline of religious belief. He noted in his writings how most people no longer believed in any afterlife or immortal soul, but how, on the contrary, our imperfect society, with its balance of nuclear terror, was all there was. This great change, Orwell mused, had to have an effect on such matters as obedience to the established law, or on the growth of crime and violence.

Again, Orwell wrote several times how he had been struck by the way the values of the liberal, capitalist, bourgeois society were being incessantly attacked by the left-wing intelligentsia and by almost the whole of the literary and artistic avant-garde. Such critics, Orwell wrote, were sawing off the branch of liberal values on which they sat. Into what morass would they—and we—fall? There was no reason to suppose that the overthrow of a democratic capitalist society would be followed automatically by a democratic socialist society. Far from it: what would follow was monstrously unknown.

This did not mean that Orwell supported British capitalism or the British Empire or the exclusiveness of his own class, from which, with a weary shrug, he had stepped out; he was never what we would call today a neoconservative. But I think that by 1947 Orwell felt that even within the intelligentsia, the recognition was dawning that there was no Utopia round the corner, that socialist measures brought no automatic end to bureaucratic malpractice and human aggression, and that the demise of the European colonial empires would bring no final solutions to the unhappiness of their subjects—in short, that liberal humanism was entering upon a period of possible exhaustion.

Not long ago it looked as if Western expansion would never cease, but recently it has seemed to falter. The great oil price hike of 1973 set off an obstinately lingering international economic recession, which has looked at times like an end of the great era of expansion. Many experts say that no swift alleviation can be expected from the current scourge of mass unemployment in the industrial countries of the West, nor any quick end to the international malaise in once-booming industries such as steelmaking and shipbuilding.

Surprising statements, such as that world-famous American corporations are no longer competitive and that giant size in industrial units is being questioned are heard today. There is talk of greater motivation of workers, and of experiments by some American firms in reshaping their employee relations along the lines of the Japanese model.

Listening to this, I thought about the economic recession in Europe and about the arrival of American intermediate-range cruise missiles in Berkshire in rural Britain. These missiles have given rise to a phenomenon, the Greenham Common women, encamped anti-nuclear protestors who will not admit men into their ranks because they regard these missiles, together with their Soviet counterparts, as symbols of male aggressiveness carried to the level of fantasy. Thinking of the blight and violence in our city centers, I have once or twice found myself wondering in my dotage whether these Greenham Common women are not right.

But enter high technology, like the good fairy. How backward was the technology of Orwell's day. Before his death, in January 1950, he may have flown in an aircraft just once, returning from Germany to England in 1945 after learning of his wife's death. Only once or twice did he watch some rudimentary television programs. Not only did he know nothing of computers of microprocessors, he also lived before the age of the transistor and the tape recorder. In his society of 1984, the act of spying on every citizen by means of television had to be carried out directly by countless other citizens.

I look today at newspaper advertisements in Orwell's Britain publicizing ever more efficient—and ever cheaper—home computers. On BBC television programs I am shown schoolboys and schoolgirls engaged in competitions in expert computer programming and design. I learn that telecommunication, with its satellites, its magic cables, and its instant transmissions, is the fastest-growing industry in the world.

How swiftly this change has come about. In the early 1950s I first heard Fritz Schumacher, the son of a distinguished German socialist, talk in London of his ideas for industrial life, which he summarized under Kropotkin's slogan "Small is beautiful." His ideas could not then have been expected to have the effect they have had, yet today, in the era of high technology, with old industries fading and new ones springing up, the trend is away from giant size. Decentralization has suddenly ceased to be a purely theoretical concept. In the successful high-technology industries some of the greatest advances have been made by small-scale enterprises. Recently I expressed my surprise to a friend in Israel that so small a country should have so successful a base of high technology. He answered that Israel could not compete in the mass manufacture of cars, but in the high technology of lasers and scanners, of missiles and aircraft, Israel could be perfectly competitive with far larger powers. And so presumably could other advanced third world countries.

It is, of course, not just high technology that will determine our future, but the use we make of it. We have all too frequently been made aware of the dangers of greater state control of citizens through high technology, against which Orwell intuitively warned in *Nineteen Eighty-Four.* As has often been said, through computers linked into the data banks of all other available computers, details of the where and when and how of the lives of individual citizens can be made known instantly to a central authority, with all the possibilities of autocratic or even despotic rule this implies.

Is there any longer a chance for the survival of privacy? Let me here be bold and say that, compared to the old-fashioned but effective tyrannies we see around us in the rest of the world, I do not think that documentation of citizens' lives by the existing data banks as yet poses any significant danger to our democratic freedoms—not if the upholders of these freedoms remain reasonably vigilant. Much more interesting is the opposite thesis, that the new computer and communication technology, if used correctly, could make for successful decentralization in the conduct of political, social, and cultural affairs. It could provide scope for much more effective action by every kind of local public authority or local voluntary citizens' group. This would be a development of which, I am sure, George Orwell would have approved.

In any event, the new era of high technology will bring with it vast social changes, both within the various national societies and between them. The mass unemployment that has troubled the Western world for the past ten years is in part a warning of the great change.

We face other crucial questions. In the Soviet Union, the most rigidly centralized society the world has ever seen, will the new technology strengthen the powers of the central Communist Politburo and the KGB to run the life of the entire vast country—or will it make for greater autonomy, greater freedom for Soviet regions, Soviet local bodies, Soviet universities, and even individual Soviet citizens? Can high technology, as already developed in India, become an instrument for overcoming the obstacles of backwardness in the rest of the third world?

In his way of life and in his political views Orwell stepped right outside his British class. Yet he felt that its culture, with its ideals of discipline, professional service, and patriotism, had been by no means entirely bad. He felt that even bourgeois hypocrisies were a crumbling barrier against the new rise of violence, as he made plain in the essay "Raffles and Miss Blandish." Orwell never himself managed to penetrate the British working class, but he carried with him an idealized picture of a "good" working-class life sustained by helpful neighbors on a local street. Such local class loyalties had of course been obliterated in the Soviet bloc. In his later years Orwell thought that they were also fading away in our Western consumer society—a society based on continuous technological change, on mass consumption and mass entertainment, and on mass politics.

What Orwell feared above all was the new collectivism of this Western society, which, in pessimistic moments, he saw as developing into a society of rootless individuals, living in the shadow of nuclear arms, entertained and talked at by slogans, fearful of the surrounding violence—and so ready for any new beliefs, however inhumane. Things have not turned out to be quite so bleak. Still, our social critics have indexed the trends in our consumer society in the most careful detail. Have these trends worked for the best in the past? Do they work for the best today? Can we believe that they will work for the best in the future? To apply ourselves to these questions is the best service we can offer to the memory of George Orwell.

Notes on Contributors

Robert Wright is a senior editor of *The Sciences,* the magazine of the New York Academy of Sciences. His column, "The Information Age," appears in that magazine regularly.

Henry Steele Commager is John Woodruff Simpson Lecturer at Amherst College.

A. E. Dick Howard is White Burkett Miller Professor of Law and Public Affairs at the University of Virginia.

O. B. Hardison, Jr., is Professor of English at Georgetown University and President of Washington Resources, Inc. His most recent book is *Entering the Maze: Identity and Change in Modern Culture.*

Eliot D. Chapple is the author of *Measuring Human Relations; Principles of Anthropology* (with C. S. Coon); *The Measure of Management* (with L. R. Sayles); *How to Supervise;* and *Rehabilitation: Dynamic of Change.*

R. Freeman Butts is a professor emeritus of Teachers College, Columbia University, and a visiting scholar at the Hoover Institution at Stanford University.

Mark E. Kann teaches political theory at the University of Southern California. His most recent books include *The American Left: Failures and Fortunes* and *The Future of American Democracy.*

Jack Golodner is Director of the Department for Professional Employees of the AFL-CIO.

David L. Sills is the Executive Associate of the Social Science Research Council in New York and the author of *The Volunteers.*

Lewis H. Lapham is the editor of *Harper's.*

S. Frederick Starr is President of Oberlin College and the author of nine books on Soviet Affairs, the most recent of which is *Red and Hot.*

Germaine Brée is Kenan Professor in the Humanities at Wake Forest University.

Mary Catherine Bateson was conducting research and teaching in Iran during most of the period 1972–78. She is now Professor of Anthropology at Amherst College.

Jan van Dorsten teaches at the University of Leiden in the Netherlands.

David Burnham is a reporter in the Washington bureau of the *New York Times* and author of *The Rise of the Computer State.*

Albert E. Gollin is Vice President of the Newspaper Advertising Bureau, where he directs research on the mass media. He was formerly President of the American Association for Public Opinion Research.

Richard M. Restak is Associate Professor of Neurology at Georgetown University Medical School and author of *The Self Seekers.*

Douglass Cater is president of Washington College in Maryland.

T. R. Fyvel was a friend of Orwell's. In 1940 he collaborated with Orwell in editing the *Searchlight Books* on war aims and in 1945 succeeded Orwell as literary editor of the socialist weekly *Tribune.* During the war Fyvel served for a time as Psychological Warfare Officer with the U.S. Fifth Army in Italy.